Strategic Development For High Technology Businesses

With Market Studies in Computers, Communications, and Computer Services

John J. Pendray
Ernest E. Keet

VPI
Value Publishing Inc.
Wilton, Connecticut

Copyright © 1987 by Value Publishing Inc., 405 Danbury Road, Wilton,
Connecticut 06897.

Library of Congress Catalog No. 87-51059
ISBN 0-9618991-0-7

About the Authors

Ernest E. Keet is a Partner of Vanguard Atlantic Ltd., an investment/merchant banking firm specializing in strategic planning and implementation of mergers and acquisitions in high technology businesses.

Mr. Keet received his Bachelor of Mechanical Engineering from Cornell University and holds a Master of Science in Operations Research from New York University. He has been President of the Software Products Group at Dun & Bradstreet, and has served as Chairman of the Software Products Committee and as Chairman of the Software Protection Committee with the Association of Data Processing Service Organizations (ADAPSO).

Mr. Keet is the author of *PREVENTING PIRACY: A Business Guide to Software Protection.*

John J. Pendray is a Partner of Vanguard Atlantic Ltd. Previously, he was Vice President, Network Engineering and Brokerage Operations at the Bunker–Ramo division of Allied Corporation.

Mr. Pendray has served as President of SESA–Honeywell Communications, Inc., President of GSI–USA, Inc., and Directeur General of Tecsi–Software in France. He was also a Captain in the United States Air Force, based at the Pentagon.

Mr. Pendray received a Bachelor of Science in Engineering Sciences from the University of Florida, and a Master of Science in Computer Science from Stanford University.

This book is dedicated to the staff and clients
of Vanguard Atlantic Ltd.,
all of whom contributed to its content.

About the Book

Preface

This book was written by the Partners and staff of Vanguard Atlantic Ltd., an investment/merchant banking firm specializing in high–tech businesses. We are especially indebted to Jan D. Rumberger who authored the market study on health care computing and to Denis B. Riney for his contributions to the study of systems integration.

Some readers may wonder why a firm specializing in mergers and acquisitions would devote so much time and talent to the analysis and development of strategic methodologies. The answer lies in our fundamental belief that successful investing requires either an opportunistic or a strategy–driven approach. Opportunism requires the skills of a horse trader and the luck of a gambler. Strategic investing requires disciplines that allow the investor to assemble pieces of a puzzle in such a way as to insure that the whole exceeds the sum of the parts.

In our experience, there is no way to transfer the skills of the horse trader, and luck benefits both opportunistic and strategy–driven investors. Opportunistic investing, however, is a zero–sum game. Because there is no change in a property's value after it is bought or sold, every winner is offset by a loser. In contrast, strategic investing results in more winners than losers, precisely to the extent that properties restructured into more efficient units have increased worth.

Our goal, then, in developing the approaches that follow, was to help our clients improve their odds of success, first by giving them a means to thoroughly understand their own businesses and their strategic opportunities, and then by allowing them to look at how specific strategic combinations could produce real value.

We have organized the chapters that follow into two distinct groups. The first four chapters develop the methodologies that are used in the balance of the book. Simply put, these method-ologies are (i) market definition and segmentation, (ii) competi-tive analysis within the immediate and adjoining segments, and (iii) assessing strategic options based on current relative position and market maturity.

Why did we develop our own methodologies? To be truthful, we did not. We simply took the best of existing approaches and welded those pieces into a new, cohesive whole. The reasons why we did not adopt classical strategic methodologies "straight out of the bottle" require a brief review of those approaches and their applicability to fast–moving businesses; that is the subject of the chapter that follows.

Because our work is heavily oriented toward the computer, com-munications, and software markets, the case studies we present are biased in these directions. We have found, however, that the techniques described can be applied successfully to any fast–moving business segment.

In conclusion, this book can be read in three distinctly different ways. First, a student of strategic theory could read Chapters 1 through 4 and then select a case study excerpt to view the theory at work.   Second, the methodologies we propose could be adopted without studying the prior art as summarized in Chapter 1. Lastly, a reader with a particular interest in turnkey data net-works, professional services, systems integration, health care computing, etc., could jump directly to those summarized market studies, referring back to Chapters 2, 3, and 4 on an as–required basis. Whichever way you approach it, we think you will find the effort to be worthwhile.

Contents

Illustrations

1

A Review of Strategic Disciplines and Theories

2 A Review of Strategic Disciplines and Theories

1

A Review of Strategic Disciplines and Theories

Introduction

The application of strategic disciplines to corporate development is a recent phenomenon; until the Second World War, debates on the relative merits of vertical and horizontal integration were at the limits of strategic theory. ("Vertical integration" refers to gaining control of the up- and down-stream sources of supply or distribution, and "horizontal integration" refers to diversifying into related businesses.)

After the War, the relatively new disciplines of operations research and management science were first applied to production and distribution problems: minimizing delivery routes, optimizing production schedules, forecasting demand, estimating rejects through statistical analyses, etc. In the last several decades, however, with these fundamental applied–mathematics problems largely solved, academic attention turned to economic theory, financial modeling, econometrics, input–output analyses, bond and option models, cash flow analyses, monetary theory, and so on. Business schools, themselves a post–war growth phenomena, quickly adopted these analytical disciplines and began to apply them to more mundane business problems, such as investment analyses, cash management models, and the evaluation of alternative business options.

Two of the three major thrusts of modern strategic theory, portfolio analysis and financial analysis, are a direct outgrowth of this effort. The third, loosely termed "shared–value planning," is really an attempt to define the elements of success in a company driven more by its culture than by top–down strategic directives. In recent years, shared–value planning (once the poor–man–out of strategic theory) has had a glorious rebirth, thanks mainly to the success of the Japanese and the perceived limitations of wholly–analytical disciplines.

Each of the theoretical approaches has tremendous value, and all three can be combined in various ways to meet individual company needs. For example, the shared–value planning approach is most suited (and the portfolio approach is least suited) to entrepreneurial businesses in high–growth markets. But we're getting ahead of the story: let's first review each approach in more detail.

Financial Analysis

This approach, the most analytical of the three, determines a value for each investment alternative and then makes strategic decisions based on their expected values. The overall strategic thrust is financial, i.e., to maximize shareholder values. Each project is evaluated as if it were a separate business which is financed solely by capital investment. The value ascribed to each undertaking is the estimated value a public market would give the business if it were separate and publicly traded, less the capital investment required. Put a different way, the value ascribed is the current value of the business to independent investors with free access to capital at competitive rates.

The theory is usually represented by a single mathematical model which discounts the cash flows to and from the business:

$$V = \sum_{n=1}^{t} \frac{C_n}{(1 + p)^n}$$

Where:

$V =$ the present value of the business

$C_n =$ net cash flow for period n

$t =$ duration of investment, expressed as a number of periods

$p =$ percent return from equivalent investments, per period

Note that p is not simply the cost of funds. It is the return expected from other equivalent investments and, therefore, must reflect both risk and possible rewards. Independent investors frequently buy shares in risky companies that pay no dividends but which promise a large future payback. Conversely, mature businesses with shares that pay generous dividends also attract capital, usually at a lower long–term cost. In either case, the share price reflects the true free–market valuation of a business in competition with other businesses for capital; exactly the rate needed to evaluate alternatives.

This discounted cash flow (DCF) model is more frequently associated with capital budgeting than with strategic planning, but the two are fundamentally related. In fact, financial analysis is the primary tool to keep the other two approaches "honest," i.e., to prevent subjective analysis from becoming quantitative gospel.

Recent renewed interest in financial analysis has led Arthur D. Little and others to propose that "Shareholder Value Creation" (SVC), an extension of the DCF approach, is a more appropriate technique. SVC adds *perceived* shareholder values to projected cash flows, both for growth in those cash flows and for the residual value of the investment. In effect, the SVC approach says that the *only* job of management is to maximize the real and perceived values of investments in relation to the cost of equity, i.e., to maximize shareholder wealth. The variables which can be controlled by management are: current returns (e.g., dividends), the cost of equity (risk–related), and perceived values (future payouts). These variables are fundamentally at odds; current payout reduces future cash flows, and higher risk increases the cost of equity but also increases the possibility of larger returns.

As SVC applies to strategic planning, all decisions are to be weighed in light of their impact on shareholder wealth. Allen H. Steed III of Arthur D. Little notes that "Shareholders are short–changed when corporations which are largely valued on one basis hold assets that are valued on another." Put another way, when a company is valued by its shareholders for its predictability of earnings, it will reduce shareholder wealth if it makes investments that increase risk in earnings (and hence the cost of equity) without offsetting attractions in perceived future value. Even if there *are* such offsets, the cost of equity may rise beyond justification if the corporation cannot appeal to a homogeneous set of investors willing to put up capital on a consistent basis. Both private and public companies must factor this into their planning; for the private company, debt and venture capital sources want different investment profiles, and for public companies "conglomeration" almost always means that the sum of the value of the business segments if set free would exceed the value placed on the whole.

An SVC approach merely discounts future real and anticipated values on a DCF basis, but with the cost of equity constantly changing to reflect an outside shareholder's valuation of the firm's cost of equity, i.e., a competitive market valuation.

Deficiencies of Financial Analysis

The most frequent error made in using analytical tools such as DCF is in mis–defining the project, investment, or business segment. For example, a company planning to enter a new business might envision a series of investments to accomplish its goal. Is this a single project? Probably not. The first acquisition or investment carries a different risk factor than does a follow-on investment. Perhaps a "seed" acquisition brings a hoped-for technology or market advantage which will then, if all goes well, be reinforced through additional internal and external investments. Note, however, that the investing company can cut-and-run between the initial investment and the subsequent investments: the investment decision is not made all at once, but serially. Accordingly, the risk premium in the follow-on investment decisions is far lower than in the initial undertaking. From a financial analysis viewpoint, the investor expects a greater premium when he buys shares in a startup than when he invests in the expansion of an existing business.

All too frequently, companies (and investors) lump their serial investment opportunities into one decision, assigning too high a risk premium to the entire venture. Professor Stewart C. Myers of the Sloan School of Management has observed that in multistage projects "The second stage is an option, and conventional discounted cash flow does not value options properly." In high-growth businesses or in businesses with substantial intangible asset value, this "option" value of the business can be considerable and, Professor Myers notes, "Standard discounted cash flow techniques will tend to understate the option value attached to growing, profitable lines of business." Clearly, an SVC approach does not solve the problem, as no single cost of equity applies to an investment with future options, and a snapshot of investor interest ignores their future opportunities to change their minds.

The critics note that there is real value in merely *being able* to reinvest in an existing business before a competitor can, and one should pay particular attention to one's "portfolio" of opportunities, not just to individual investment decisions.

Portfolio Analysis

Portfolio analysis attempts to rate the relative positions of existing and planned business units according to their competitive position or strengths and the market attractiveness. The Boston Consulting Group's approach rests on the contention that high market share produces high margins. Arthur D. Little's approach contends that profits are determined by market position and industry maturity. McKinsey & Company states that the business's strength and the industry's attractiveness are the determining factors. G.E. adopted an approach that merged some of the BCG growth–share matrix with McKinsey's approach. And so on. In each case a graphical method is used to position one business segment versus another.

The BCG theory predicts that high market share will yield high profits, primarily due to the greater experience and reduced unit costs which will result from producing higher volumes and gaining greater experience than competitors. As a result, the theory states, there are natural equilibriums that all markets will achieve as the various players change position. In the words of Bruce D. Henderson, one of the fathers of portfolio theory:

> "A stable competitive environment never has more than three significant competitors, the largest of which never has more than four times the market share of the smallest....All competitors wishing to survive will have to grow faster than the market in order to maintain their relative market shares with fewer competitors. The eventual losers will have increasingly negative cash flows if they try to grow at all."

In the visual presentation of this theory, business segments are positioned on a "growth share matrix." The horizontal axis of this matrix corresponds to the market share of a business compared to its major competitor; the vertical axis to the annual rate of growth of the market. Because the theory states that the decline in costs accruing from accumulated experience is logarithmic, the horizontal axis is usually plotted on a logarithmic scale, reflecting the theoretical link between market share and accumulated volume (Figure 1.1). The dividing line between "high" and "low" relative market share is usually placed between 1.0 and 1.5.

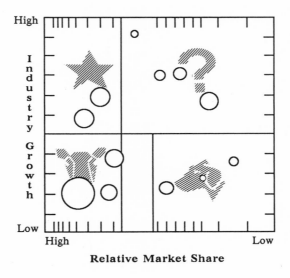

Relative Market Share

Figure 1.1

High–market–share businesses in low–growth markets are termed "cash cows" because they throw off excess cash (their cost of production and their need for further investment to gain market share being low). High–market–share businesses in high–growth markets are "stars" that produce attractive profits with minimal additional investment. Conversely, low–market–share companies are cash eaters. Low–market–share companies in high–growth markets are "question marks" that can, with significant investment, become "stars" or, if left to languish, be the victims of a consolidating market. Low–market–share businesses in low–

growth markets are the ultimate losers: "dogs" that consume cash with no hope of returns.

Another graphical device frequently used to illustrate the BCG portfolio analysis theory plots relative market share on the horizontal axis and after–tax return on sales on the vertical axis. Once again, the horizontal axis is logarithmic to reflect the link between accumulated volume, costs, market share, and hence profitability (Figure 1.2). Supporters of the BCG theory suggest that in "equilibrium" business profitability will fall within a "normative band" on this graph, i.e., the after–tax returns of any business will be ultimately determined by its relative market share.

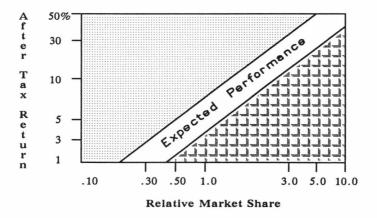

Figure 1.2

To achieve desired growth rates and profit objectives for the business as a whole, the BCG portfolio analysis approach suggests that companies use the excess cash generated by "cash cows," and the proceeds from the sale of dogs ("to whom?", you might ask, since the theory never advises *buying* dogs), to nurture the growth of the "question marks" into "star" status.   As markets mature, the theory adds, the "stars" become "cash cows" (and, if left unattended, the "question marks" become "dogs"). For continued growth and profitability, the portfolio must be constantly changing, with a flow of high–growth but low–share businesses into the higher–share quadrant.

McKinsey & Company noted that the BCG approach failed to take into account the relative attractiveness of different industries, and offered a substitute approach that only indirectly considered market share. Their "nine box matrix" (Figure 1.3) positions the portfolio units by business strength and industry attractiveness.

Industry Attractiveness				
Low	Medium	High	High	Business Strength
		Invest/ Grow		
	Selectivity/ Earnings		Medium	
Harvest/ Divest			Low	

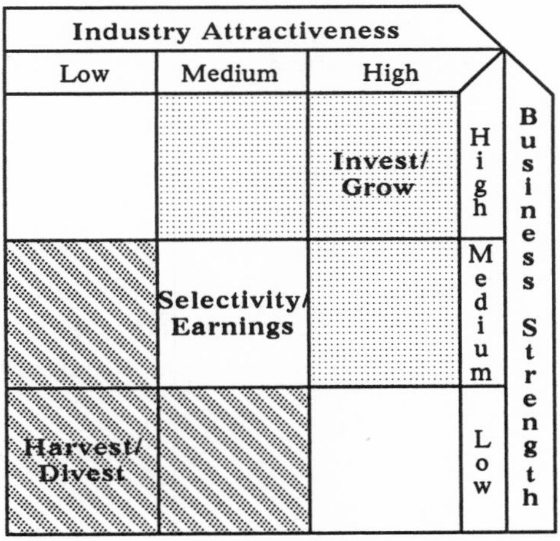

Figure 1.3

General Electric attempted to combine the approaches in a "nine box matrix" that illustrated both the industry size and the business units' relative market share within that industry as "bubbles" on the McKinsey matrix (Figure 1.4).

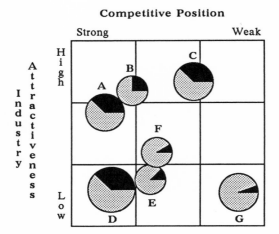

Figure 1.4

Regardless of the technique used, a clear separation of each business unit from the rest of the "portfolio" is required, as is a clear understanding of the market segment in which the unit competes. Once these data are gathered, the relative businesses and business opportunities can, the theories state, be ranked against each other to determine which business units should be "milked," grown, or divested.

Deficiencies of Portfolio Analysis

Portfolio analysis has the simplicity and elegance of most basic theories. But, as Galileo and, later, Einstein demonstrated, basic theories may accurately explain observed data for the wrong reasons. Markets are not static, capital is generally available from the capital markets (as well as from the "cash cows"), high returns on investment *are* possible with small market share, most companies do not (and cannot) run as a portfolio of separated businesses, markets are not homogeneous (a "cash cow" in the U.S. might be a "star" in Spain), and cost reduction does not always come with volume and experience.

Professor Arnoldo C. Hax of MIT's Alfred P. Sloan School of Management even suggests that return on investment *declines* temporarily as market share increases (Figure 1.5). Professor Hax notes that while large market share does allow for the exploitation of the experience curve, small market share focused on a unique competitive differentiation can be highly profitable.

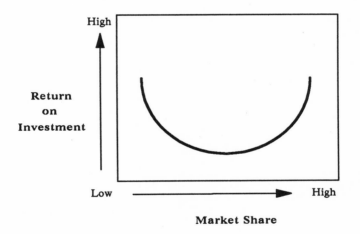

Figure 1.5

The place *not* to be is at the bottom of the U–curve, "With no cost advantage and no distinctive product to offer."

Many also criticize portfolio analysis because it relies on the semi–subjective definition of markets. If "data processing" is a market, then all but IBM will be also–rans. If, however, "mainframe application software for life insurance companies" is a market, new entrants will concern themselves far more with Cybertek and Continuum than they will with IBM. The definition of markets is, in fact, critical to *all* strategic planning and analysis, and is the subject we will address in detail in Chapter 2. Portfolio analysis, however, requires *precise* market definitions or else the theory turns on itself: golden opportunities can be made to look like "dogs" if a fast–growth sub–market is lumped into a larger but slower–moving market segment; major threats can be ignored if the evolution of currently–separate markets into a single competitive arena is missed.

Another criticism of portfolio analysis comes from the advocates of Shareholder Value Creation (SVC) who, as noted above, claim that business conglomeration may "balance" the portfolio and diversify the risk, but will also drive away "pure play" investors willing to provide low–cost equity. Utility company investors, they note, invest in, and pay a premium for, "cash cows" because of the current income (dividends) they provide. High–rollers invest in, and pay a premium for, "question marks" and "stars" that may not pay dividends but that hold immense future promise (and risk). To assume that all cash must flow *within* the portfolio and not from these outside investors is unrealistic, they note, and the "balanced" portfolio will almost surely make equity more expensive.

Perhaps the strongest criticism of portfolio analysis is that the goal of a "balanced" portfolio may be wrong. Professor William W. Alberts of the University of Washington along with James M. Taggart of Marakon Associates have suggested that "Effective capital investment planning tells us that their application should not depend on whether a particular unit is independent or a member of a portfolio." They point out that an "unbalanced" portfolio (i.e., one that generates or consumes cash) may be desirable. In fact, "Prospectively profitable portfolios can be balanced or unbalanced, and balanced portfolios can be prospectively profitable or unprofitable...a company should not add cash consuming units to its portfolio because it doesn't want to pay dividends; it should add cash consuming units because they will be valuable and thus create value."

While the portfolio analysis approach to strategic planning has immense value, especially in understanding the business elements and opportunities, its biggest drawback may be that it demands a top–down and totally analytical approach to planning. Counter-logical as it might seem, some of our greatest business successes have adopted bottom–up and culture–driven planning, some-times called "shared–value planning."

Shared–Value Planning

Relying on insight and long–term development of technological competitiveness, shared–value planning revolves around a set of cultural values which are commonly shared throughout an enter-prise. While market share is of concern, the dominant factor in shared–value planning is this shared set of beliefs. Professor Steven C. Wheelwright of Stanford University has itemized the differences between portfolio–based and shared–value planning (Figure 1.6) and has contrasted two firms that practice each at the extreme: Texas Instruments (portfolio based) and Hewlett–Packard (shared–value based).

PORTFOLIO   VS   SHARED-VALUE

Process =	Rational & analytical Comprehensive & explicit	Logical & incremental Prioritized & implicit
Trade-offs =	Strategy driven	Value driven
Decisions =	Top down	Evolve
Results =	Strategic change	Past performance
Key to Success =	Structure	Collective management

Figure 1.6

Professor Wheelright observes that:

"Texas Instruments (TI) prefers a long-term, low-cost market position" whereas Hewlett Packard (HP) "seeks competitive advantages in selected smaller markets... HP tends to create new markets, but then exits as cost-driven competitors enter...TI emphasizes continual price cuts to parallel cost reductions in order to build volume and take advantage of shared experience...HP holds prices longer...HP concentrates on flexible production... TI concentrates on cost effective production...TI looks for a portfolio that includes low-growth businesses with dominant market shares to provide cash...HP is likely to want all high-growth businesses with dominant market

shares...it is harder at HP to make dramatic shifts in resource commitments."

It is obvious that shared–value planning is not a discipline at all, but rather an approach that emphasizes top–down articulation of a corporate image, style, culture, and purpose with bottom–up development of specific projects, business segments, and markets. Shared–value planning techniques emphasize communication and consensual analysis methodologies for planning rather than dogmatic approaches to strategic decision making.

We've looked at the pluses and minuses of portfolio analysis; what are the deficiencies of shared–value planning?

Deficiencies of Shared–Value Planning

Whereas the portfolio approach concentrates too much on analyses of markets and market position, the shared–value planning approach concentrates too much on past successes, R&D, and "gut feel." Too much power is delegated and, as people tend to protect the status quo, shared–value planning frequently prevents rapid strategic moves into new high–technology businesses. Shared–value planning also commits too many resources to fading businesses, preventing bold responses to competitive moves or longer–term planned repositioning. Nonetheless, the Hewlett–Packards, 3Ms, and Japan Inc. have all demonstrated the benefits of accumulated *personnel* experience in contrast to (but not necessarily in competition with) the benefits of accumulated *production* experience.

Merging the Approaches: Non–dogmatic Planning

Strategic planning, despite the many advances in analytical theory and applied mathematics, remains an art. The tools and disciplines used must fit the company; forcing a portfolio analysis approach on a decentralized "family" of individual performers with shared values could destroy the business. Similarly, delegating decision making and resource allocation in a centralized and hierarchical organization can result in chaos and waste. The use of financial analysis absent a longer–term commitment to how follow–on investments will be treated is a loose cannon on the deck. Each method has its place and all three can complement one another (see Figure 1.7), but no one technique is the answer.

Merging the Approaches

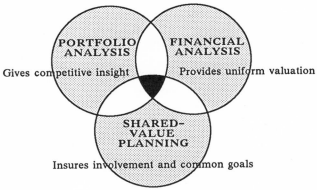

Figure 1.7

The Japanese have been (overly?) praised for their single–purposed, consensus management, shared–value approach to markets or products. Yet the Japanese have not relied solely on shared–value planning. In fact, the Japanese are the foremost practitioners of market development based on experience and volume–induced cost reduction. One could almost think of the Japanese nation as a "portfolio," the individual companies being carefully selected business units within the portfolio and the planning process within each of these business units being value based.

Similarly, any business should select elements of portfolio and financial analysis and of shared–value planning in the development of a strategic plan.

Portfolio analysis applies best to a competitive analysis of the business's options, a means to give weight to future options which cannot be valued through financial analysis or which might be overlooked in shared–value planning. Financial analysis, on the other hand, provides a uniform means of valuing investment alternatives which, if properly applied, can highlight the subjective assumptions in either portfolio analysis or shared–value planning. Shared–value planning, in turn, insures the involvement of the organization, at all levels, in the planning process. It is a style of management, not a set of disciplines and techniques, designed to foster independent management thinking and common goals. All three approaches should be used in the development of a strategic plan to give it the best chance of success.

2

The Crucial First Step:
Market Definition

2

The Crucial First Step: Market Definition

Introduction

As we have seen, any strategic approach requires a careful knowledge of markets, and market definition is not as obvious as it might first sound. Is there a "computer services industry" for example? We think not, any more than there is a "manufacturing industry." Manufacturing is an economic sector, like agriculture. Aerospace manufacturing relates to automobile manufacturing as growing cotton relates to growing wheat: they share common resources and tools, but that's about all.

Two frequently mentioned "industries"-- computer services and communications--are really collections of business segments with very different characteristics, e.g., airline reservations and credit reporting, sharing common resources and tools (computers and communication links), but that's about all.

According to Professor Derek F. Abell of Harvard University[1], there are three components to defining a market (see Figure 2.1):

1. Customers served (who),
2. Functions provided (what), and
3. Technologies employed (how).

Generally, the customers served will be defined by both economic sector (e.g., manufacturing) and industrial segment (e.g.,

[1]See Professor Abell's *Defining the Business*, Prentice-Hall, Inc., 1980.

farm equipment). Functions provided generally will define the
competitive arena for a product or a service (e.g., inventory control), while technologies employed will reflect major differences
in approach (e.g., micro versus mainframe software).

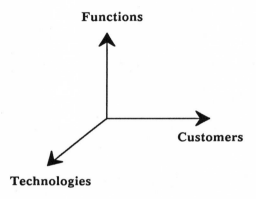

Figure 2.1

Within the services sector, "computer services" defines a group
of computer–based technologies comprised of many market segments (e.g., mainframe software, custom contract development).
Similarly, functions performed are frequently grouped into major
categories, such as "development" or "operations," which also
comprise multiple market segments.   Customers are typically
classified by economic sector and industry, e.g., automotive
manufacturing, but in some cases must be defined by the actual
user/purchaser (e.g., automotive engineers).

Defining a market in these three dimensions yields meaningful
results, for example, mainframe software (technology) performing computer–aided design (function) for automotive engineers
(customer), whereas leaving out any one dimension would define

too broad a market for a meaningful competitive analysis (mainframe computer–aided design software would include architectural design, electronic design, mechanical design, etc.).

Markets are usually competitive in all three dimensions. Automotive engineers buy a lot more than design software, and at the "customer" level the sellers of automated drafting systems may be defining their market identically (and be competing for the same limited budget monies) as the software vendor. Similarly, the technologies available (mainframe software, micro software, timesharing, turnkey systems, and workstations) can define differing approaches to the same customer and function. Lastly, to the extent that a product or service performs a commonly-needed (i.e., "cross industry") function, it may serve several markets (both aerospace and automotive engineers may be buyers of automated drafting systems).

Market Dynamics

Market dynamics are frequently created by movements along one or more of the definitional axes. It is not useful to say "we are the market leaders in mainframe computer–aided design systems for automotive engineers" if automotive engineers are switching to individual workstations in droves. This was the error made by many timesharing vendors in the '70s who properly identified their markets by customer and function, but who failed to recognize the threat of two new technologies (primarily software packages for in–house use and microcomputers).

Strategic plans that neglect the intersections of markets are doomed to failure. Markets can intersect in several ways. First, so–called "cross industry" products can impact markets defined solely (and erroneously) by the customers served. An example would be the profound impact of generalized (i.e., cross–industry) accounting packages on vendors of accounting packages to specific industrial (i.e., "vertical") markets, such as the insurance industry. Second, markets can intersect where different technologies with changing price/performance characteristics come together. Just as discount airlines have redefined the inter-city transportation market, so have microcomputers permanently redefined nearly every computer or communications service market with a technology component. An example, of course, would be every time-shared application that did not need the burdens and power of a large host computer, a centralized database, or a multi-terminal network.

Another example would be the market frequently described as "professional services," i.e., the provision of expert manpower to do customer-unique work (Figure 2.2). A computer services market exists for "design and project management" (function) for software development (technology) in data processing departments (customer). Another market exists for the same function and technology in user departments, frequently in direct competition with the in-house data processing department. Yet a third market exists for "design and project management" (function) for audit and accounting systems (technology) in finance departments (customer). The second differs from the first because the customers are different (user department needs are notably different from a data processing department's). The third varies from the second because the technologies (audit and accounting versus software development) are different.

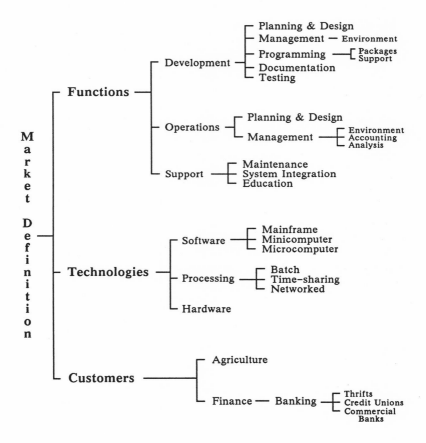

Figure 2.2

Classical market share analysis would neglect the potential, but not yet realized, ability of these markets to overlap. Big–Eight accounting firms would be considered a competitor for project management contracts in finance and accounting, but might not be considered (erroneously) a threat in the other two markets where their existing share is small.

In a three-dimensional analysis, however, the key question is how difficult is it to expand on any axis. Accounting and audit firms already have solid positions in finance and audit departments, are experienced in audit and accounting technologies (and increasingly in data processing technologies), and have begun to compete aggressively for more "management services" business (i.e., design and project management). As the technological leap becomes shorter, so will the temptation for these firms to compete more outside of their traditional markets.

Markets must be viewed, therefore, from two perspectives: (i) what is the existing position? and (ii) how could that position change or be changed? We will return to examine these questions in more depth, i.e., what are market segments that adjoin the segment we are interested in, and what are the barriers to "tunneling" between segments, in the chapter that follows.

3

Applying Strategic Theory
to High Technology Businesses

3

Applying Strategic Theory to High Technology Businesses

Introduction

As noted in the earlier presentation of strategic disciplines and theories, none of the modern analytical techniques provides sufficient guidance in fast–moving industries. In fact, the more rapidly changing the markets or technologies, the more imprecise become analyses based on quantitative long–term projections. As a result, most planning in high–technology businesses is based more on experience, "feel," technological issues, and human factors (e.g., the tendency to back previous winners), than on quantitative analysis. Out of this soup will generally evolve some home–grown but structured form of shared–value planning based on collective experience, consensus on the relative merits of competing opportunities and technologies, and individual advocacy.

Can the virtues of quantitative analysis be applied to strategic planning in rapidly–changing businesses? If so, can they coexist with the shared–value analyses that have been so successful in many entrepreneurial firms? We think the answer to both questions is "yes."

Quantitative Analysis for Fast–moving Businesses

Modifications to the more static forms of quantitative analysis can increase their usefulness in planning for fast–moving businesses. Some of the theories of portfolio analysis, for example the belief that accumulated experience will reduce costs, are less meaningful to businesses undergoing constant change.   Other principles have great value, however.

Consider the issue of relative market share. Is it true that a competitor with two times the market share has an advantage in a changing market?   Clearly, but not only for reasons of accumulated experience. Simply put, the larger competitor can match, dollar for dollar, the advertising, direct selling, development, support, etc., of the smaller competitor and still have a larger remaining profit. This fact becomes even more critical as the intangible (e.g., intellectual property) content of a product increases in relation to its manufactured or purchased components. In a "pure" intellectual property sale, such as in the licensing of software products, only the direct sales and support costs (mainly personnel–related) vary with volume; all other costs give the larger competitor a significant advantage.

By spending more absolutely than the smaller player, the larger competitor can continue to report equivalent margins while increasing market share. The *only* chance the smaller competitor has in such an environment is to spend available monies more efficiently, i.e., to develop superior products for the same or lower amounts, to advertise more effectively at the same or lower cost, etc. In a truly "efficient" environment, this disequilibrium would be impossible; as superior performance is based totally on (i) better personnel or management or (ii) capital investment, the larger competitor would simply hire away the superior performers or invest more.

This is actually what happens, but with a twist. The larger competitor spends available monies on something that creates a marketplace differentiation in products or services and the smaller competitor is forced to reply in kind, e.g., by developing a similar capability. This keeps the smaller competitor's profits at a minimum and restricts the flow of investment capital, frequently the death knell for high–tech companies.

How can we capture this concept of relative market share among direct competitors within a segment?

Competitive Share Analysis

A simple modification to the growth share matrix and a careful market segment definition can help answer this question.   By displaying the relative competitive position of directly competitive players versus their growth rates relative to the market's rate of growth, a snapshot of relative opportunities emerges (Figure 3.1). To illustrate hidden strength and staying power, each competitor is plotted as a circle, the circle's size representing total applicable company strength (not just volume in the market segment being reviewed).

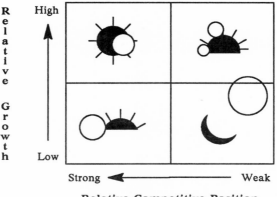

Figure 3.1

"Applicable strength" is difficult to define, but should include current revenues in the relevant business segment *plus* additional revenues from related segments, the total representing the resources a competitor could directly apply to this segment, if he so chose. Note that as the attractiveness of the segment increases, so does the incentive for a large player to shift resources from less attractive areas, as shareholder value will increase to the degree that future reward expectations offset current expenditures. If the area is of critical strategic importance to a competitor, e.g., office systems for IBM, then a great deal of the resources of the corporation must be considered as being applicable in that segment. Of course, resources cannot be moved overnight, and hence there is another consideration: how easily a player can "tunnel" from one market into another, a subject to be explored shortly.

As all the plotted competitors play in the same market segment, high growth rates with low market share signify not "question marks" but "rising suns," companies with a competitive advantage who appear to be defying the laws of gravity and gaining market share. From the perspective of the potential investor, these companies are, in the case of high–tech industries, the most attractive; they have something, usually a superior technology (remember, invention creates significant market disequilibriums), that allows them to prosper in the face of superior force. The high–market–share, high–growth companies are (again, temporarily) unbeatable. They are well managed and have good enough products and services to stay with or ahead of the competition, but at significantly higher volumes. They are, to continue our naming convention, "noonday suns."

The high–market–share companies that are growing more slowly than the market are simply abandoning their future in this segment. They most likely acquired their lofty market position through an innovative technology, or through early entry into the market, or both. Their failure to use this position to continue to innovate or sell or service may shortly cost them dearly. They are "setting suns" that are doomed to be classified as "waning moons" if not rescued soon. Frequently these "setting suns" bring in new top management who spend extra monies playing catch–up which, in turn, lowers profits and drives away investors. The lowered valuation then attracts savvy buyers who either smell a turnaround or value the customer base.

Finally, the low–share companies that consistently trail the market's growth rate are either mismanaged or underfinanced (usually both). Whether startups or collapsed "setting suns," these companies have an almost impossible task: they must spend large sums to rapidly grow in a high–growth market. Why bother?

From a planning standpoint, this competitive position matrix can truly help companies evaluate their existing position and their investment opportunities. If you fall in the "setting sun" quadrant, the message should be clear: change management and invest quickly while shielding yourself from takeover until depressed profits are behind you. Perhaps look to acquire a small "rising sun" in a form of reverse–merger (you bring the money and the market, they supply everything else). If you are already the "rising sun," do the reverse: quickly look for the funding to grow rapidly or go buy a "setting sun" before a "setting sun" or a "noonday sun" buys you. If you are a "waning or a waxing moon," find a business broker to help you find one of the people that P. T. Barnum said was born every minute.

"Noonday suns" can be the most vulnerable of the lot, in terms of the risks they face. Highly valued by insiders and outsiders, they can develop a misplaced feeling of invincibility, a belief that their market segment will grow forever, and an inflated company valuation. As a result, diversification into other market segments can come too late. When the market's growth begins to slow, as every "hot" market will, they will slow with it. Because they were good managers in good times does not assure that they will handle this slowdown properly. The key question should be: as top dog in this market segment, what *other* market segments are they developing before their prime market slows down?

Note: An example of the calculations used for developing a relative competitive position matrix is given in Chapter 9.

The Impact of Market Maturity

The strategic decision–making process must therefore be based not only on a business's relative performance and competitive position, but also on the maturity of the market. Figure 3.2 is a simplified visual representation of the logical decision alternatives based on these three considerations.

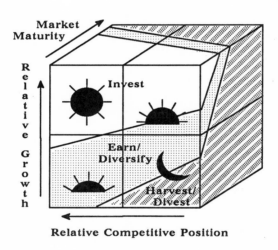

Figure 3.2

Companies with good relative performance and high market share (the "noonday suns") should invest heavily if the market is young and emerging. As the market matures, these same companies should shift emphasis to earnings and diversification into younger markets which they have a better chance of dominating. "Rising suns" should begin the earn/diversify process even earlier in the market cycle, while their unsustainable performance can be used to attract funding for the transition. "Setting suns" have some hope of generating earnings and diversifying early in the market's cycle but have no option but to harvest the business once the market is established. "Waning moons" must take what they can get and run with it, unless they can find the immense resources required to become "rising suns" through unbelievably rapid growth.

For the outside investor looking to buy into a new market or technology, the "noonday suns" are almost certainly a bad buy once a market has some maturity. They are then fully (over?) priced, by both their investors and their management, and they are probably arrogant as well. The best buys are "rising suns" and "noonday suns" early in a market's development. Investments in both "rising suns" and "setting suns" can create hybrid "noonday suns," a low-cost road to riches.

Note that cash flows within a "portfolio" of business units is irrelevant to this analysis. During the early development of a market segment, every player is investing either to gain share or to diversify. When the market growth rate is high, even if it has begun to decline, *all* of the companies on the matrix are cash-eaters and the availability of investment capital (a deep-pockets owner, savvy venture capital, public investors, etc.) must be assured for all but the "waxing or waning moons."

All of the foregoing analyses assume a finite set of players in a market; this is clearly insufficient. As markets evolve, new players appear on the sidelines, companies in related but more mature markets with sufficient capital and the same need to diversify that we discussed above. Because of their strengths in related markets, these unannounced competitors can enter a market swiftly, doing severe damage to the other players in the process. In high-technology industries there are many strong players who wait for markets to emerge before they move with strength; IBM's whole business history appears to follow this path.

How can we anticipate these new players with the *potential* to be market-share leaders in the future? For example, whenever a market gets to be of sufficient size (usually around $500 million),

IBM simply powers their way to dominance, not necessarily with superior products or services. IBM's very presence drives away customers and capital, making the natural crowding–out process even more efficient.

Even if market analysis does not uncover an IBM as being a significant existing competitor, it and other powerhouse players in related industries must be considered in any strategic planning. To do so, the definition of related businesses becomes a critical factor, as it is from these related businesses that "tunneling" is the easiest.

The "Tunneling Effect"

In Chapter 2 we noted that markets are best defined on three axes: customers served (who), functions provided (what), and technologies employed (how). Existing players who have significant overlapping market positions on any one of these three–dimensional axes can easily "tunnel" into adjoining markets, whereas an outsider with nothing more than cash at his disposal would face greater obstacles to entry.

Let's consider a specific example. In computer–aided design, there are many different technologies (specialized workstations, mainframe software products, micro–based software products, etc.), myriad functions (electrical engineering, solid modeling, mechanical design, robotic simulation, etc.), and many customer sets (VLSI design engineers, architects, etc.).

A position on any one of these three axes can require a substantial investment. Difficult technologies must be mastered, these technologies must then be functionally applied to develop unique solutions to a market segment's needs, and finally a set of customers must be created and nurtured. Once a technology has been mastered, however, its possible application in other markets reduces the barrier to entry into that market by one dimension. If both the technology and customer set are the same, for example, only the reapplication of that technology to other functions stands as an obstacle to new markets.

In the CAD example, a vendor of stand–alone workstations to automotive engineers for use in structural design (Figure 3.3) might discover that some of the functions (topographical analysis, rotation of three–dimensional forms, etc.) had equal applicability in related markets (e.g., for architects). In this example, the vendor could "tunnel" into the related market by expending monies on (i) application of the existing technologies to different functions and (ii) developing a new customer set.

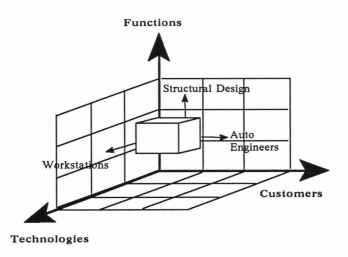

Figure 3.3

Similarly, while this workstation vendor was tunneling into the architectural markets, a vendor of general–purpose microcomputer software might be tunneling into the market for automotive structural design. From a strategic planning standpoint then, the question is not merely how to define a market, but how to define the adjoining markets. Like a chess game where a tree structure can define the options n moves out, the consideration of entry barriers can be extended to players n steps away from the market under consideration, in all three dimensions. Inversely, a search for new markets should always start with the current market, and then testing adjoining markets in a step–wise fashion. The barriers to entering a new market increase geometrically with the distance from an existing market.

A simple graphical device for examining adjoining markets is to map the competitive position matrix into three similar matrices which represent competitive analyses on a single market–definition axis only (Figure 3.4).

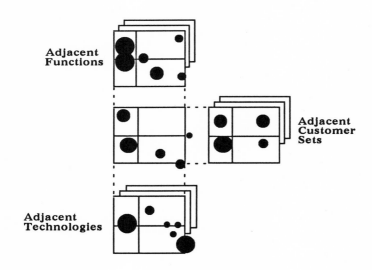

Figure 3.4

One describes the relative competitive position of all players currently possessing similar technologies. A second describes the position of all suppliers to a given set of customers. The third positions all players offering a given set of functions.

Consider another example. Utilities must produce detailed analyses to support their rate applications. As the complexity of these analyses has increased, there has been a concomitant increase in the use of computers and software. If we were to examine the market for (i) financial analysis and (ii) utilities using (iii) software packages, we would find a niche market with a handful of players. If we were to examine adjoining markets on each axis, however, we would see that the function (financial analysis) is performed in many other market segments and that there are many well-established players with either products or services who could easily tunnel into the utility market.

Similarly, if we examine the customers, we find many suppliers with strong positions in utilities, for example the public accounting firms, who could quickly offer a tailored rate-case analysis product or service to their existing customers. Lastly, if we look at the established software suppliers, we will find many with generalized decision support or financial analysis offerings who could easily customize their products to fit this market niche (IBM included).

Mapping the three dimensions in this fashion can lead to paranoia ("Will IBM or The Big Eight jump into my market?"), but it can also lead to the development of rational plans for protecting an existing market while tunneling into others.

The Value of Related Markets

For the smaller competitor, "customers served" is the most important of the three market definition axes. To the extent that customers form a "fraternity," i.e., they are a homogeneous group with good inter-group communications (trade associations, specialized journals, etc.), there is special value in providing them with multiple products. Thus, supplying one market segment for a given set of customers can reinforce another; cost-sharing between segments can improve profits, and vendor loyalties can be a strong additional barrier to competitive entry. This is not to say that inadequate technology or an inability to provide required functions can be overcome by having multiple products for a single set of customers.

Suppliers with high fixed costs and large generalized distribution channels find moves into new customer markets using existing technologies to be more attractive than attempting to build specialized services for these customers. Their goal is to sell bulk communications facilities, computers, software packages, supplies, services, etc., whenever the accumulated volumes from *multiple* customer sets bring the cost–reducing synergies so essential to price–sensitive (i.e., commodity) markets.

There is virtually no need to fear such a large–volume supplier in a specialty market. As previously noted, IBM usually targets a market with an annual sales volume of over $500 million and, when it does, it soon dominates it. For suppliers of cross–industry products, the early days of a market's development can be deceptive since volume is not sufficient to attract the truly large player who needs massive volume to justify moving his large resources into a new market. As soon as the smaller players have established sufficient volume to warrant such a move, however, the larger players move with a vengeance.

Even good–sized companies (e.g., Hewlett–Packard) have integrated this phenomenon into their planning, abandoning markets they helped create just when the total market volume reaches levels attractive to commodity suppliers. Many other companies have, to their ultimate dismay, held on tenaciously to a cross–industry market for too long, having failed to recognize that the *only* strategy in such markets that works is to "hit–and–run," abandoning the field of battle when the stronger competitors enter it (i.e., when price and volume become significant factors).

In the *cross–industry* segments of high–technology markets, we believe this hit–and–run strategy to be the correct one for nearly every participant; IBM, the Japanese, and possibly AT&T excluded. For the small– and medium–sized companies it is essential. In the long run, for all but the very–large players, customer–

specific markets are preferable to technology–specific or functional markets. If an existing cross–industry market position is technology– or function–specific, the appropriate strategy is to select, "tunnel" into, and reinforce customer–specific vertical markets. Then, once a vertical market position has been established, related markets can be added by supplying additional technologies or functions to the same customers. This provides new high–margin opportunities while erecting strong barriers to other entry.

Only after a vertical (customer–specific) set of markets has been established with dominant technologies and multiple functionality should these technologies or markets be used to tunnel into other vertical markets.

4

The Impact of Market Maturity:
A More Detailed Look

4

The Impact of Market Maturity:
A More Detailed Look

In Chapter 3 we explored the application of strategic theory to high technology industries. The impact of market maturity was portrayed by the cube reproduced here as Figure 4.1.

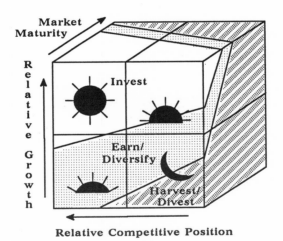

Figure 4.1

What we didn't discuss was how one knows where on the maturity scale a market segment is at any point in time. In this discussion, we will use some of the same concepts to develop techniques for estimating the maturity of a market segment within its life cycle.

The best single indicator of market maturity is the rate of growth of the segment.   Since new markets are small, it is easy to have impressive growth rates when expressed, as is usual, in percentage increase of sales from period to period.   Mature markets are large relative to their average sizes throughout their life cycles, and growth rates are, consequently, harder to come by, even though the sales dollars involved may be much larger than for young markets.   This leads us to suspect that market size within the life cycle is another indicator of maturity, and is somewhat the inverse of growth rates.   How do we resolve the apparent problem of two indicators which seem to be interdependent?

Let's philosophize for a moment about what usually causes a market to be mature, i.e., stop growing.   The first thing that comes to mind is saturation.

While this is certainly a valid indicator for some markets, it is seldom the cause for market slowdowns in high technology segments.   Technological displacement is a more important factor in these businesses.   For example, the market for time–sharing-based financial modeling didn't saturate; it got kicked in the teeth by PC–based spreadsheets like Lotus 1–2–3.

Also, changes occurring on the other axes, functions performed and users served, can have significant impact on the growth rate of a business.   For example, adding new functions or new users can provide growth for a technologically–based business by tunneling into adjacent segments.   This gives us a hint that our three–dimensional market definition approach may also be useful in gauging the maturity of a market segment since threats to, and opportunities for, growth usually come from other nearby segments.

We'll define a new term, market sector, to encompass more than one segment. A market sector can be thought of as a segment and its nearby segments.   In the three–dimensional representation, there are 26 contiguous segments around any particular segment, forming a Rubik's cube of adjacent markets.   Adding the

next surrounding layer will bring another 98 segments into consideration, and so on. With this precise image in mind, we'll now
muddy the issue by adding the word "relevant" to complete the
definition of a market sector: *a market sector* is composed of a
central segment and those nearby segments relevant to it. For
example, a technology–based enterprise looking for growth
would consider nearby segments containing new functions and/or
users as relevant for analysis, but would look to nearby segments
with different technologies providing similar functions to his user
base as being relevant for a threat analysis. Only some of the
nearby segments are "relevant" for a given analysis; therefore,
knowing which ones to include in the sector requires some experience and, perhaps, several iterations.

So far we have said:

(i) market segment maturity is indicated by two related
variables: growth rate and relative market size;

(ii) market segment maturation is frequently a result of
the impact from events in nearby market segments; and,

(iii) a market segment and its relevant nearby segments is
defined as a market sector.

In order to develop a representational methodology for the ex-
pression of this concept, we will now introduce the Market Sector
Growth–Size Matrix (Figure 4.2).

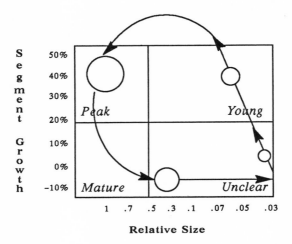

Figure 4.2

The Y axis represents market growth rate, and the X axis is the relative size of market segment within a market sector, normalized to the largest segment in the sector. The individual market segments are positioned on this matrix according to their growth rate and relative size within the sector, using a bubble which is proportional to the sales of the segment.

Looking at Figure 4.2, we see a typical life cycle for a segment within its sector. At the early stage, the segment is small, its growth is experimental and low, and its future is unclear. As the segment succeeds, its growth accelerates, and it gains in size by taking sales from other segments within the sector (thereby increasing its relative size), and represents a young market segment. With time, the segment reaches its maximum growth rate, then its maximum relative size (where it isn't necessarily the biggest segment in the sector) and is at its peak. As new segments emerge within a sector and begin to capture sales, the older ones see their growth stagnate and the segment is mature.

Eventually, the growth rate becomes negative, the market size begins to shrink, and the segment's future is again unclear as it struggles for revival or, more likely, passes from the scene. Not all segments will follow the same trajectory, and not all sectors will use the same growth rates on the Y axis; however, the form and direction should be preserved in most cases.

One last comment on our understood third variable: time. Like all trajectories, the velocity is not constant. In general, the right side of the matrix is traversed quickly in high-technology markets with newcomers rising swiftly and old-timers making fast exits. The peak and mature phases tend to be longer lasting (thank goodness for the investors) as the pure inertia of large size resists change.

To demonstrate these concepts on a simple case study, let's return to the market for financial modeling mentioned earlier. In Figure 4.3a, we see the market sector for interactive computer-based financial modeling circa 1975, which was the heyday of time-sharing services. By 1980, we had seen the entry of "spreadsheet" programs on microcomputers, and Visicalc was growing rapidly, as shown in Figure 4.3b. Lotus 1-2-3 began taking share in 1982, pushing Visicalc into the mature quadrant and sealing the coffin on time-sharing financial modeling (Figure 4.3c). Today, Lotus 1-2-3 is thoroughly ensconced in the peak quadrant, with no threats of significance yet identified (Figure 4.3d).

**The
Market
Sector
for
Interactive
Computer-
based
Financial
Modeling**

S
e
g
m
e
n
t

G
r
o
w
t
h

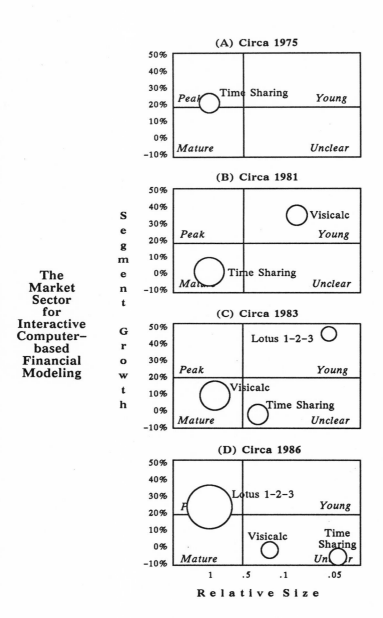

Relative Size

Figure 4.3

By neglecting other players like Multiplan, Jazz, etc., we made this well known case study a simple one, and, therefore, a good example. Most cases are more complex, as we'll see when we apply this technique in our analysis of the market sector for intelligent multiplexors (Chapter 9).

5

Market Study #1:
Turnkey Data Networks

5

Market Study #1: Turnkey Data Networks

Overview

The first four chapters of this book have discussed the theory and practice of strategy development for high technology industries. The time has come to put the theories described into practice through a real–life market study. There will be several more such studies to follow, but the first may be the most illustrative, because data communications in general and the market for turnkey data networks in particular are typical of rapidly changing business segments.

As computer and communications technologies merge, one wonders if all data communications will disappear into the Integrated Services Digital Network (ISDN) in such short order that bothering to segment this market as it exists would be a waste of time. It is true that communication switches are now computers and that many computers function mainly as data communication switches; however, the integration of voice and data (let alone video) over common media will be a long time coming. We expect that the mid–1990s will see the deployment of true ISDNs as more digital media become available, e.g., fiber optics, and when integrated switching technology is ubiquitous. Until then, the significant investment in installed analog facilities and copper wire will continue to require investment in devices to convert digital information into analog signals and back again. This cost alone retards the rush toward digitizing everything so it can be ISDNed.

Even more telling is the old saw that the usability of an emerging technology is inversely proportional to the amount of attention it receives in the trade press. By this measure, ISDN is a long way off, since the press is loudly trumpeting its praises. Any way you look at it, it seems the data communications market will be with us as a distinct market for enough years to justify segmentation, analysis, and investment.

For openers, let's take a look at the market for turnkey data networks. By turnkey data network, we mean everything necessary for end–to–end data transport between many terminals and host processors in a widely dispersed area, excepting (usually) the actual transmission media. Turnkey data networks, therefore, include the hardware, software, and services necessary for line termination, concentration, protocol conversion, distribution and transmission switching, network management, technical monitoring, accounting, and (hopefully) security control.

We should point out that this case study will limit itself to data networks, excluding the market for integrated networks even though they usually contain a significant data network portion. While the market for integrated networks clearly offers opportunities for suppliers of data networks, these vendors are normally subcontractors to large systems integrators like Electronic Data Systems, Computer Sciences Corporation, Boeing Computer Services, and American Telephone & Telegraph. Leading–edge communications users such as large governments and Fortune 500 companies have recently started requesting integrated communications networks (with voice being the dominant part), and we expect this trend to continue. This is an exciting new area; however, for purposes of this discussion, we will limit our analysis to the market for stand–alone data networks. This turnkey data network market is fairly mature and, consequently, useful as an example for market segmentation.

Applying the methods presented in Chapters 2, 3, and 4, we will search for an understanding of the market segments by analyzing the three axes: functions provided, technology employed, and customers served. Then, logical combinations of these three elements will be made to produce market segments.

Data Network Functions

The functions performed by turnkey data networks start, and end, with line termination. Physical connections of terminals and hosts may be as simple as plugging in a direct digital cable or as complicated as installing a modem–based long–distance line. For hosts, the network connection unit may interface to, or, in some cases, assume the function of, a front–end processor.

Concentration is one of the most important functions of data networks, since it provides most of the economic gains. Simply put, concentration is the process of consolidating low–speed and/or sparsely–used data streams onto highly utilized (thus, more economic) facilities.   In this context, speed conversion of data streams is a subfunction of concentration. (We'll return to the subject of economic gains at the end of this section.)

Conversion of protocols is another function of the data network. Often, this is accomplished in the concentrator or front–end, but may reside in the terminal, host, or any intermediate device. This is a software–intensive task and has been the main impediment to true multi–connected networking.

Switching of data is clearly the fundamental ingredient in data networking, i.e., being able to send data from any device on the network to any other.

The functions discussed so far are necessary, and sufficient, to qualify as a network, and we will refer to these as the "basic functions." Any data network of respectable size also requires some, if not all, of the functions which follow, which we will call the "management functions."

Controlling the access, use, load distribution, fallback and recovery, and configuration of a network is representative of the tasks included in elementary network management. Technical control provides diagnostics and measurements of all the physical resources of the network, e.g., modems, lines, concentrators, and switches.

Accounting, statistics, and accountability are important functions in large–scale networks, be they for public or private use. It is necessary to know who used what resource, when, and in what manner. This information is useful for accurate cost determination for billing, and crucial for managing the changing resource requirements found in most networks.

Last, and often least, are security measures. There are three main areas of security: network access control, transmission integrity, and data access. The first two are properly the responsibility of the network, the third being a task for the host containing the data. Network access control is the management of who gets on the the network and what rights they have to use network resources. Preserving transmission integrity includes not only guaranteeing delivery of error–free data, but also protecting data from unauthorized interception during transmission.

Like just about everything in communications, turnkey data networks have been affected by deregulation of the telephone companies. Laying aside the difficult–to–measure advantages of interconnectivity, resource sharing, etc., the bare economics of data networking have undergone a fundamental, and still changing, metamorphosis. In the "good old days," long–distance transmission costs were artificially high in order to cross–subsidize local distribution costs which were artificially low. Great financial savings could be gained by increasing the utilization of transmission lines through networks which provided data concentration at the distribution hubs. With deregulation, transmission costs are plummeting and distribution costs are skyrocketing, thus reducing the advantage of concentrating for transmission. There is a lot of scurrying around by data network suppliers to find solutions to the distribution problem in order to base the economic gains on reduction of distribution costs. (This issue will recur as a key factor in later sections of this analysis.)

One other impact of deregulation has been advantageous to the vendors of turnkey data networks: loss of the single–vendor telecommunications supplier. Now that large telecommunications users are faced with the responsibility of choosing and integrating communications services to optimize cost/performance, the value of network management is high. Although it's tough to put a dollar amount on it, the management functions that used to be considered as just a nice "plus" are becoming indispensable.

The Technology

Multiplexing is the oldest, and most prevalent, technology used for data communications. The main function performed by multiplexors is concentration of data to increase circuit utilization; however, today's intelligent multiplexors can also provide some switching, speed and code conversion, protocol interfacing, etc. Where a small number of densely populated sites need to send significant amounts of data between points, the efficiency of the multiplexor is hard to beat, as is evidenced in the hot market for T–1 (1.544 mbps) multiplexors. But when you get into the meshing of many widely dispersed terminals and hosts, the dominant functions become those of switching, protocol conversion, and network management. Multiplexing technology doesn't perform these functions very well.

Data PBXs (private branch exchanges) are another technology touted for data communications networking, and they certainly have their place. Wherever large numbers of data circuits need to switch with each other, the data PBX should give a good showing. This is typical of a densely populated local area requirement, i.e., a distribution function, and the data PBX is often a good low–cost alternative. However, a full–function networking technology, it isn't.

Packet switching, where a data message is cut into small "packets" which are each routed over the network, is the only technology to offer all of the functionality desired, and it didn't just appear on the scene. It's been developing for about twenty years and has finally settled into two forms: IBM's Systems Network Architecture (SNA) and the–rest–of–the–world's X.25. Although SNA's functionality extends far beyond the functions considered here, the approaches in the areas under consideration are quite different between SNA and X.25. Even the struggle between these two is almost over as most X.25 network suppliers can interface SNA to X.25. (It went mostly unnoticed, but IBM was an active participant in the X.25 effort and made major contributions to its definition and development.) The emerging ISDN standards will probably include X.25 (or its internetworking cousin, X.75) packet switching for data transmission. We'll be very surprised if SNA is included in ISDN, so even more protocol converting between IBM's approach and international standards is on the horizon.

In summary, there are really only two technologies in turnkey data networking: intelligent multiplexors and packet switching. They both compete for networks requiring only basic functions, and, in this segment, multiplexing usually has a cost advantage. For large networks needing the management functions, there's only one viable alternative to consider: packet switching. Moreover, packet switching appears to be fairly free from competitive threat for the immediate future because of its probable inclusion in the ISDN standards.

Figure 5.1 depicts which functions are provided by the different technologies.

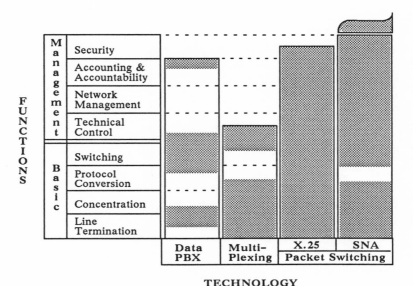

TECHNOLOGY

Figure 5.1

Customers

The three types of customers are public data network (PDN) operators, large organizations, e.g., Fortune 500 companies, trade associations, etc., and government. Let's examine them in reverse order.

The federal government is clearly a big user of data networks, and will continue to be so for the forseeable future. Many requirements exist for basic-function networks, and the multiplexor vendors are very active in this segment. The GSA catalog contains most of the components manufactured for these networks, and tough price competition is normal.

Large-scale networks are fewer in number, but really large! The DOD and intelligence agencies' networks are among the largest in the world. The government market for large turnkey data networks is fairly–well penetrated, with most growth coming from additions to installed networks.

Some activity has been seen in state and large local governments, and we expect this to continue at a not–very–exciting rate. There just isn't enough critical need in our decentralized form of government to support a lot of big data networks, so this segment tends toward smaller networks of limited geographical scope.

Large organizations have been another mainstay of the market. In particular, many organizations have needs which are met by a small network providing basic functions, and many multiplexor networks have been installed. While we think the overall growth rate for small networks has peaked in this segment, there should continue to be respectable business from both the installed base and some new systems as costs decrease and higher speed circuits, e.g., T–1, become generally available.

For several years now, market researchers have been predicting impressive growth figures for full–function data networks in large organizations. Alas, after the international companies and financial institutions, the second layer of this market has failed to materialize. We suspect that a combination of three factors is responsible for this. In order, they are: the uncertainty created by deregulation, the dominant position of IBM in most of the subject organizations, and the lack of enough heterogeneity and dispersal to require full–function data networking. In fact, the truth is probably that this market segment is quite large, growing, and belongs to IBM. Even so, we are still optimistic that rapid growth for X.25 packet networks is hiding in this customer base somewhere, but we have no idea when the elements will unite to produce it.

Public data network (PDN) operators may be another kettle of fish. The 1984–85 period saw a lot of low–revenue positioning by large numbers of strong communications companies. Most of the Bell Operating Companies (BOCs) went through the RFP cycle, selected a vendor, and installed a node here, a three-node network there. The large independent telephone companies installed packet nets for internal use, with an ability to use them as a base for a PDN. IBM offered its new entry into the data services business through the SNA-based Information Network. MCI, Federal Express, DHL, etc., installed initial packet networks for electronic document transmission. In brief, an impressive number of pawns were placed, which hasn't produced much revenue for the vendors, yet.

Our prediction for this market segment for the next several years is very bullish, even though this may be another year of treading water while the BOCs sort out the final effects of deregulation. In fact, we expect that the BOCs will finally provide the needed boost to really get this market going and move toward ISDN.

Our confidence is based on a simple analysis. All that local-loop wire is still the best answer to the distribution problem for most of the world, and the telephone companies are finding some impressive ways to leverage this asset. For example, the data/voice multiplexing (DVM) being introduced with the new Centrex (a PBX-like service provided from the central-office exchange) allows data to be sent simultaneously with voice over the local loop. The telephone company digital central office just strips off the data stream and routes it to a co-resident packet switch which then sends it on its merry way over concentrated high speed data circuits, or back to the Centrex user over another local loop. In this manner, the Centrex provides local area networking and access to long-haul internetworking at the same time. When true ISDN arrives, the central office may do all these functions in the same switch (which will be a bigger computer).

The main thing to notice about all this is that no new major facilities were required, only a central-office-based packet switch and some DOV (data over voice) devices for the local loops. Since we're using the old copper local loops which are already there, the transmission media are free! They already know how to drive these circuits at 256 kbps, so there is plenty of bandwidth. Voila!, the distribution cost problem just went away, and not only for the big users, but for everyone with a telephone.   As this technology matures and advances, we expect data transmission to be a very competitive offering from your local telephone company.   Since the threat of bypass is spurring the BOCs on, this may be one of the first beneficial products of deregulation.

Figure 5.2 summarizes the functional requirements of the different categories of customers.

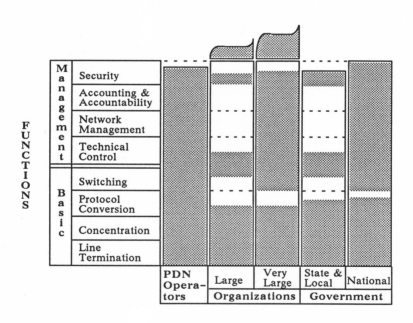

CUSTOMERS

Figure 5.2

Market Segmentation

Now that we have identified the functions, technologies, and customers, we're ready to combine them to define market segments. While we do this, we'll also introduce some of the key players and do some competitive analyses. Let's start with an elaboration on the technology issue, which will lead into some conclusions about the government market.

Thanks to accepted standards and hard–won experience, building a network which satisfies the basic set of functions is fairly straightforward. Users can choose from a wide range of competitive products, even mixing vendors in most cases. Such is not the case when full functionality is required, because providing management functions on an end–to–end basis for a heterogeneous network is a bear of a job. Not only are all these functions software intensive, worse yet, each function requires some software in each and every hardware component of the network, and the whole thing has to work as an integrated system! For this reason, the number of suppliers capable of providing full–function turnkey data networks is limited, and, because vendors' network management systems are incompatible, each customer tends to become a captive of his vendor. The degree of difficulty is very high, and entry into this market is expensive and difficult.

We said earlier that the government market for basic–function multiplexor networks is fairly mature, with aggressive price competition between the established vendors. Infotron, Paradyne, Racal–Milgo, Timeplex, and Codex are some of the main players in this segment and will continue to grab the lion's share by building on their installed base.

In full–function networking, governments have been nice enough to be the patrons of packet switching and have funded the projects which produced most of the technology and vendors. In the U.S., ARPANET was the first packet network and provided the basis for Bolt Beranek and Newman, Inc.'s (BBN) entry into the

business as both a direct supplier and as the founding sponsor of
Telenet (now merged into the joint venture of GTE and United
Telecom).   M/A–COM's first generation of equipment was for
Telenet, but the new generation of CP 9000 was funded by a
large government agency.   Northern Telecom, from Canada,
SESA, from France, Siemens, from Germany, and so forth, have
all been launched in the business by projects from national gov-
ernments, usually their own.   Only a few companies, such as
Tymnet and Telematics, have boot–strapped their way into the
business.

As customers, governments tend to stick with the same suppliers.
Even in the open–competition U.S., BBN has succeeded in
maintaining a near–monopoly on Department of Defense (DOD)
business (to their credit and against strong competition).   In con-
sequence, the customer base of national governments, while being sizable, is pretty well sewn up, and we don't see this as a
high–growth opportunity area for non–ensconced vendors.
There will be some new openings, but the price will be high and
the probability of winning will be low.   This is good news for
BBN, who dominates the DOD market, but not so good for others who may try to enter.   Other suppliers will certainly continue
to have government business of meaningful volume, but the days
of new entrants finding government funding for development of
new systems is probably over.   *Conclusion*:   this segment of the
market for turnkey data networks is showing signs of being a mature market belonging to the existing players.

Large organizations are probably the best sector for basic function data networks over the next several years.   Deregulation is
forcing users to shoulder the burden of managing and optimizing
their communications networks.   We anticipate that many
smaller organizations who depended on the telephone company
for total service in the past will install multiplexor data networks
to reduce costs.   Since the full networking functionality of packet
networks is usually not required for organizations having only
several geographical sites, SNA and X.25 packet networks are

rarely justified. The growth in this segment should be sufficient to provide increasing revenues for those vendors previously listed in the government section, as well as to support some less-well-known network suppliers such as Case Rixon, Digital Communications Associates, Comdesign, Tellabs, and Teltone. This is not meant to be an exhaustive list of vendors, rather it is illustrative of the number of players which this segment will support in the near future. The shakeout of this market is not likely until 1988, or later.

As was discussed earlier, the market growth for full-function data networks in the very large organizations has been disappointing for all but IBM. We expect IBM to continue to prosper among its true-blue customer base as it fights a rear-guard action against universal X.25 networking. Compatible manufacturers such as NCR Comten and Amdahl will also ride this wave successfully, but be more aggressive in their support of X.25. Sooner or later (it's probably already later), the desire to interconnect beyond the IBM-architectured world should open this market segment for X.25 data networks, but it isn't happening yet to any great extent. In fact, there may never be enough growth here to support the optimistic market study projections for open networks, because IBM has clearly defined communications as a strategic thrust and they are good at protecting their strategic interests. Nevertheless, one new entrant deserves mention: Telematics. By adopting a different strategy, this young company has positioned itself rather well within this segment. They have become an original equipment manufacturer (OEM) to other suppliers of communications equipment and large users like the General Electric Information Services Company. In doing this, they have accelerated the process of software development by dealing with software-savvy organizations.

The last customer set, PDN operators, will, in our opinion, be the most exciting one in the 1987–89 period. It will be dominated by the BOCs who will emerge from deregulation as aggressive competitors in every communications-based business. Since PDN operators see the communications world as a very large heterogeneous network which they wish to manage, they are natural purchasers of X.25 packet data networks and they see data transmission as an area of new opportunity. Multiplexing, as a technology, is commonly used by BOCs for line concentration, but we don't see it being used much for data networks.

If we're right, all the pawn positioning of the last few years will produce a very active game as deregulation gets untangled and the BOCs get untied. Low-cost Centrex data facilities could bring packet data networking to almost everyone, and that's a big market! So let's take a brief look at some of the suppliers to BOCs.

In spite of Telenet's and Tymnet's early position in the U.S. packet network business, the telephone companies have mainly gone to two foreign-owned classical telephone equipment suppliers for their initial networks. Siemens and Northern Telecom (in that order) have won the overwhelming majority of the business. This isn't too surprising, since both Telenet and Tymnet have PDN operations as their main business and are, therefore, competitors. BBN is in a few PDNs. M/A COM had a joint relationship with United Telecom (which is now dead due to the GTE-United Telecom joint venture). There's an occasional other system here and there, but mostly it's Siemens and Northern Telecom. The revenues from these sales haven't amounted to much yet, but if it takes off as we expect, the growth on these small installed bases will be impressive. If it doesn't catch on here, it may never catch on at all and the whole major market for X.25 networks predicted for so many years will have been a mirage. We're not ready to believe that, yet.

Figures 5.3, 5.4, and 5.5 portray the intersections of the main technologies with the other two dimensions of functions and customers. The next logical step is to analyze each segment for its size and growth characteristics over the fourth dimension, time. This step is left as a recreational exercise for the reader!

INTELLIGENT MULTIPLEXOR
DATA NETWORK SEGMENTS

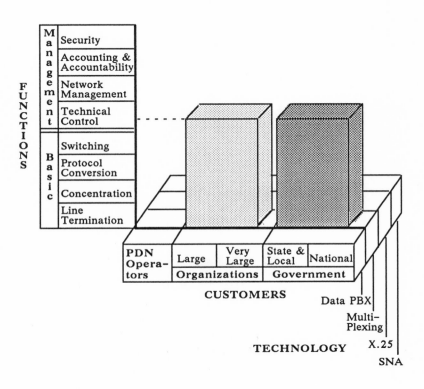

Figure 5.3

SNA
DATA NETWORK SEGMENTS

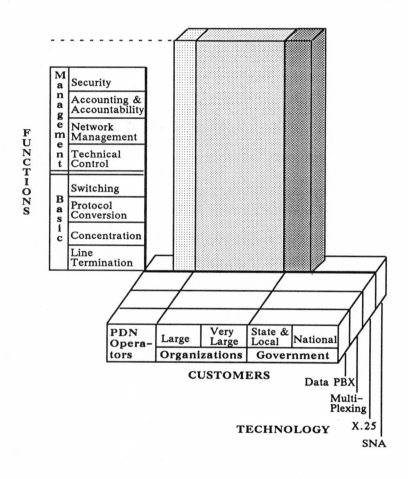

Figure 5.4

X.25 PACKET SWITCHING
DATA NETWORK SEGMENTS

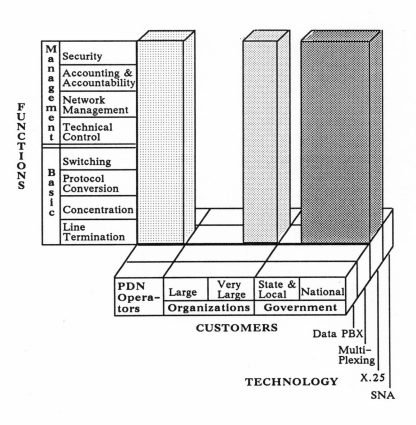

Figure 5.5

Competitive Analysis of Packet–switched Turnkey Data Network Markets

We have concluded that three main markets exist: intelligent multiplexor networks, SNA networks, and X.25 packet–switched networks. We will now apply our strategic analysis methodology to the market for packet–switched networks. Figure 5.6 is a three–axis top–down summary representation of the communications market, highlighting the components of packet–switched networks. Figure 5.7 is a three–dimensional representation of the segments within the packet–switched turnkey data networks market.

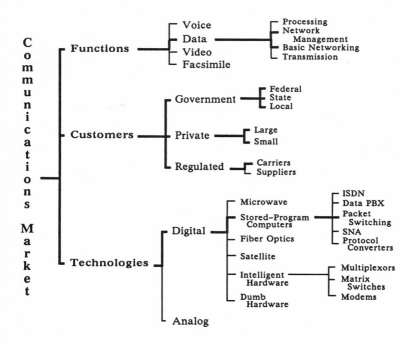

Figure 5.6

X.25 PACKET SWITCHING
DATA NETWORK SEGMENTS

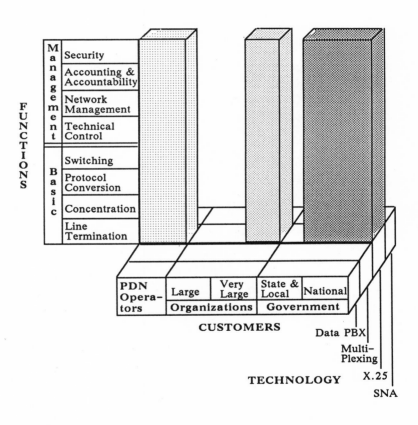

Figure 5.7

A Portfolio Analysis

First, let's take a look at a "portfolio analysis" for one of the players in this market. We'll pick Bolt Beranek and Newman, Inc., (BBN) for several reasons: they are a competitor in all three segments of this market, this market represents a significant part of BBN's total business, they are a darling of the stock market, they are a strategically managed company, and, most importantly, they hold up very well under close competitive scrutiny.

Our growth–share matrix presents BBN's activities as shown in Figure 5.8. They are dominant in the federal government segment; significant, and rising, in the large–organization segment; and small, but growing very fast, in the BOC segment.

**TURNKEY DATA NETWORKS
PACKET SWITCHED**

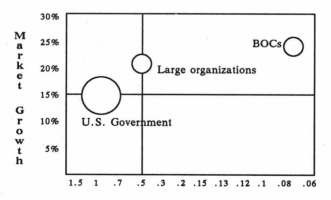

Relative Market Share

Figure 5.8

In portfolio analysis parlance, BBN has, respectively, a star becoming a cash cow, a question mark becoming a star, and a young question mark. Portfolio analysis theory would say to reinvest the profits from the maturing government segment into the high–growth question mark of the BOC segment, while continuing investment in the large–organizations segment. Keeping these conclusions in mind, let's do the competitive analysis.

Relative Competitive Position

Our estimates of segment growth rates and applicable strengths for the various players in the three segments yield the competitive analysis matrices shown in Figures 5.9, 5.10, and 5.11. Based on these competitive analyses, we would say that BBN has a high "rising sun" in the large–organizations segment, a very desirable "noonday sun" in the government segment, and a weak "waning moon" in the BOC segment. Concentrating investments in the large–organizations segment would be encouraged, while further investment in the BOC segment doesn't appear wise: a direct contradiction to the portfolio analysis result.

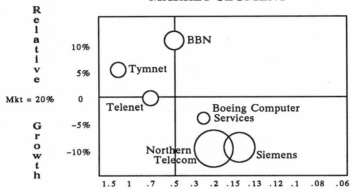

Figure 5.9

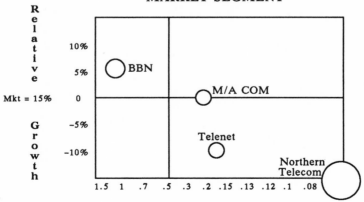

Figure 5.10

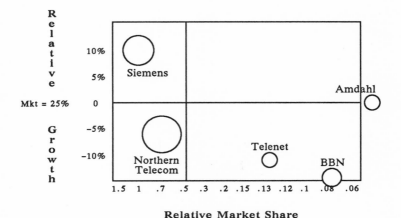

Figure 5.11

The Impact of Market Maturity

Estimating market direction is probably one of the riskiest games in town, and we don't intend to wager heavily. We'll simply speculate that the government segment is fairly mature, while the segments for both large organizations and BOCs are relatively young. Expressed in terms of our market maturity cube, BBN should earn/diversify in the government segment, invest in the large–organizations segment, and harvest/divest in the BOC segment.

Tunneling Threat Analysis

Since we're so unexcited about the BOC segment opportunity for BBN, let's use it as the focus for our tunneling threat analysis. Figures 5.12 and 5.13 show two scenarios in which we gaze up the functions axis, down the users axis, and out the technologies axis. Although we haven't labeled the players in the adjoining matrices, it shouldn't require much effort to identify the large applicable strengths of Northern Telecom, Siemens, AT&T, and IBM. One has to conclude that the market, in general, and the BOC segment, in particular, is of strategic interest to some impressively threatening players. Once again, the BOC segment seems to be one to avoid for BBN.

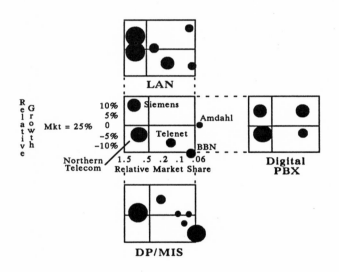

Figure 5.12

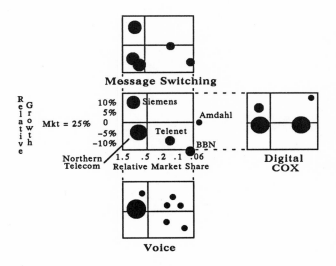

Figure 5.13

Also, BBN's opportunity to tunnel into adjacent segments seems to be limited by the large size and interests of the dominant players in those segments. Investment in the segment for large organizations looks like the best bet.

Summary

The methodology we have used provides a good framework for analysis of this fast-moving market. If the estimates for the example are anywhere near reality, a clear and consistent strategy for BBN emerges.

6

Market Study #2:
Health Care Computing Services

6

Market Study #2:
Health Care Computing Services

Overview

Is the health care computing market sick? While it may be difficult to prove that health care computing is "ill," this market is
showing some of the symptoms of old age. For example, many
of the market leaders in hospital information systems, having en-
joyed rapid growth for the past ten years, are beginning to exhibit
growth and profitability problems. In light of this, many have
turned to product and company acquisitions as their primary
growth strategy. We have watched player after player be gobbled
up in the past few years (e.g., Amherst, Mediflex, and Computer
Resources Inc., by HBO; Compucare, J.S. Data, and
Stoneybrook Systems, by Baxter Travenol; and many more).

Our definitional view of these markets is pictured in Figure 6.1.
To illustrate, there is a market segment for turnkey systems
(technology), encompassing diagnostic/testing and monitoring instruments to automate laboratories (functions), in investor-
owned hospitals (customers). One company in this segment is
Cybermedic.

Another segment is packaged software (technology), for patient
accounting (function), at state and local government–run psychi-
atric hospitals (customers). A player in this segment is Advanced
Computing Techniques.

A third, and one of the larger segments (at least historically), is
shared network services (technology), to automate financial sys-

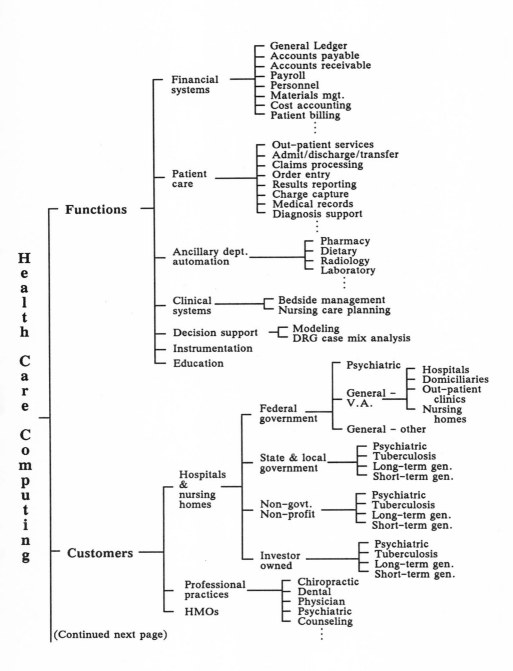

(Continued next page)

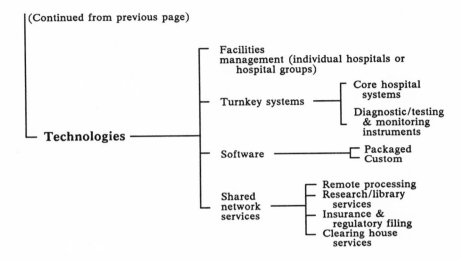

Figure 6.1

tems and patient care (functions), in non–government and investor–owned hospitals (customers).   Here we find big players like McDonnell Douglas Information Systems and Shared Medical Systems.

A final example is the segment which provides financial and patient care automation (functions) to professional medical practices (customers), via packaged software (technology).   Players here include Cycare Systems and McDonnell Douglas Physicians Systems.

While these are true examples, it is also true that each of the companies mentioned above participates in several market segments.   These companies, as well as many others, have followed one or more of the four classical tactics for market expansion:

1. Develop or acquire more functions for sale to existing customers using an existing technology base.

2. Develop or acquire new customers, often by modifying existing products to perform similar functions required by the *new* target customers (e.g., Cycare Systems' expansion into HMO's market from a good position in professional practices).

3. Move existing functions to new technologies. This often involves strategic alliances and product repackaging. A good example would be Keane Associates' recent partnering with IBM.

4. "Tunnel" into an entirely different, nearby, market (e.g., insurance systems). It is difficult to find examples of firms who are tunneling *out* of health care computing, but rather easy to find examples of firms tunneling *in* (e.g., Hospital Corporation of America, an owner and manager of hospitals, recently purchased 20% of Cycare; Baxter Travenol has expanded from hospital supplies into the systems market).

Most of the market leaders have been employing *all* of the first three tactics, while trying to grab short–term market share by beating the competition *within* the segment. (This can be particularly fierce when the segment is maturing). The more mature a segment becomes, the more you see price wars, blitz campaigns, giveaways, and downsizing (usually in an attempt to reach the remaining but less desirable potential customer). This is not just a condition particular to health care computing, but, quite arguably, is the condition of the entire U.S. health care industry.

One obvious cause is the size of the available market. The American Hospital Association identifies about 7,000 hospitals in the U.S., the vast majority being small (under 300 beds). It has been this size for years and now even appears to be declining. This eliminates the majority of hospitals as prospects for large, expensive systems (usually mainframe–based, with prices in excess of $500,000). One researcher estimates that market saturation in large–scale hospital financial systems is already near the 90% level. Consequently, it is safe to conclude that the patient's condition is chronic, especially for the mainframe–based solutions vendors. This does not imply, however, that there are no opportunity areas (at least in the short–term).

The health care computing industry got a shot in the arm a few years ago with the TEFRA legislation and the introduction of DRGs (Diagnostic Related Groupings). These brought the concept of "cost control" to hospital management (as opposed to the "cost–plus" concept of the past). They also initiated a more rapid shift toward out–patient care, home health care, and other alternative treatment approaches. As a result, health care systems products are changing rapidly.

Computing technology changes have also had a big effect on health care market segments. Micro computers have been rapidly accepted in hospitals: as workstations, turnkey departmental products, and as embedded components in testing and diagnostic equipment.

Relative Competitive Positions

Coupled with a more experienced and sophisticated set of buyers, this has caused a significant shift away from remote shared processing, toward in–house (including facilities management) solutions. In fact, in the last two years, the smaller firms, offering comprehensive, integrated solutions on minis and micros, have done quite well in relation to the old guard mainframe and shared systems companies. To illustrate, examine the following relative competitive position changes for the first four years of this decade (Figure 6.2). This data includes both hardware and software sales. However, when we add 1985 and 1986 figures, more sweeping changes will become evident. For example, HBO, which looked formidable in the 1980–1984 period, has been losing ground lately. We expect to see Shared Medical and McDonnell Douglas lose more market share and several new players appear, thanks mainly to the rapid decline of the shared services in favor of packaged software, turnkey systems, and facilities management.

Also, with maturity forcing vendors to go after smaller hospitals to secure new clients, vendors like Systems Associates Inc., Infostat, Keane, and Dynamic Control (targeting the smaller hospital) have been gaining ground, especially in relation to firms such as Compucare, IBM, Technicon, Datacare, and Unisys.

RELATIVE COMPETITIVE POSITION
(Based on sales 1980 – 1984)

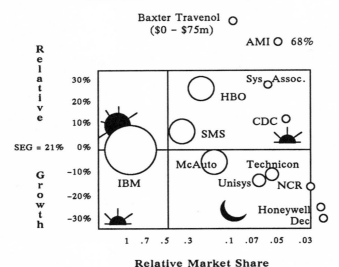

Relative Market Share

Figure 6.2

Figure 6.2 still hasn't achieved proper, detailed, segmentation. Drawing conclusions and basing strategy on this high a level of detail can prove dangerous. While we obviously cannot address all of the many segments of this marketplace, we would like to examine one in detail for the purposes of illustration: the professional practices customer. In addition to its larger absolute population relative to hospitals (160,000 versus 7,000), the facts clearly show that physician practices are becoming much larger.

All categories are growing, with the exception of solo practices, and the largest growth is in the 8–15 and over 50 member groups. With this shift toward larger practices, and the potential blurring of distinctions between professional practices and HMOs, more and more computing resources are moving into the practices. This is negatively affecting the remote and shared processing segments (the technology of choice for the smaller group practice) and positively affecting the packaged software, turnkey systems, and facilities management vendors.

The other key seems to be the current emphasis on "connecting" doctors to hospitals, doctors to other doctors, hospitals to satellite hospitals, and everyone to the insurance carriers. Take, for example, Systems Associates' announcement that they have formed an alliance with Curtis 1000 Information Systems. Their new system allows physicians, processing locally, to link directly with SAI's "larger" hospital systems in order to obtain information on patients and make decisions on medical care. (And this is a vendor whose target customer is hospitals *under* 300 beds).

Figure 6.3 shows the relative competitive position (from 1983 through 1985) of players who service the professional practice segments.

RELATIVE COMPETITIVE POSITION

Physician Practices Segment (1983 – 1985)

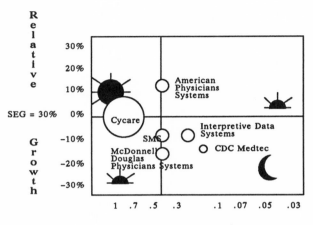

Relative Market Share

Figure 6.3

The entire annual market for this segment is about $500 million and should continue to grow in excess of 20% for several years. Most of the market leaders are approaching this segment from different starting points and with different strengths. For example, Cycare comes from a base in professional practice software and turnkey systems, but also now provides shared services and facilities management. It is also planning to tunnel into the larger hospital systems segment, (e.g., Cycare's announcement of a deal with Carraway for their hospital information system rights). SMS, on the other hand, started in hospital systems and is moving into professional practices by focusing on practices connected to hospitals they currently serve.

The other dimension is geographic distribution. Due to the wide dispersion of medical practices, there are a large number of regional providers. Some 600 vendors can be identified at the regional and local level. It is obvious that only a large firm with broad geographic dispersion can hope to dominate this segment. This would suggest a strategy of acquiring many small companies in geographically-targeted areas, a technique not uncommon to other computer services markets (the growth of ADP followed this pattern). It is also possible that a very large player which already has the required geographic coverage (even if it is not currently in health care computing) could move fast to capture a significant position.

Tunneling Threats

Tunneling from one segment to another has been prevalent for some time within this industry. Consequently, most of the industry leaders claim to have product offerings for all four delivery systems (micro, mini, mainframe, and time-sharing) across all broad customer types. Tunneling into health care computing segments has already taken place: by companies which were (and are) customers (e.g., hospital management companies); and by firms providing different services to the same customers (e.g., hospital supplies). An example of the former is Hospital Corporation of America; the latter, Baxter Travenol/American Hospital Supply.

Where could new competition come from in the future? Certainly, more hospitals (and hospital groups) could enter the market, either via acquisition, minority investment, or by attempting to sell systems they have developed internally.

TUNNELING THREATS

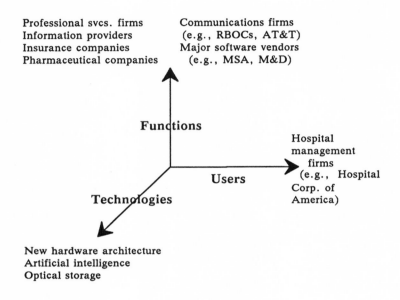

Professional svcs. firms
Information providers
Insurance companies
Pharmaceutical companies

Communications firms
(e.g., RBOCs, AT&T)
Major software vendors
(e.g., MSA, M&D)

Functions

Hospital
management
firms
(e.g., Hospital
Corp. of
America)

Users

Technologies

New hardware architecture
Artificial intelligence
Optical storage

Figure 6.4

Looking down the functions axis (Figure 6.4), a case could also be made for entry by pharmaceutical firms, communications companies (e.g., AT&T or the Regional Bell Operating Companies – RBOCs), insurance companies, professional services vendors, or even information providers. (For example, Dun & Bradstreet, currently is a third–party claims processor through Plan Services, a software vendor for claims and benefit systems through Erisco, a cross–industry financial applications vendor through McCormack & Dodge, with a formidable delivery vehicle for on–line database services).

While a case could be made for *any* of these types of companies becoming interested in health care computing, we would watch the RBOCs closely. They have a strong desire to expand their current businesses, are becoming more national in scope, are gaining experience in systems integration involving intensive communications, are financially strong, and, although still somewhat regional, already have a vendor relationship with health care customers and insurance carriers for voice communications, computer equipment servicing, software, and micro product distribution.

While theoretically possible, we suspect hardware vendors will not directly move into health care computing. What we *do* expect to see is new types of "machines" being used (e.g., expert systems on specialized hardware, or utilization of optical storage devices, bar code systems, etc.,) in health care applications.

Although we have focused our attention on the U.S. markets, the number of hospitals in the rest of the world is an eye-opener. For example, the Soviets have almost four times as many hospitals as the U.S. and the Chinese have ten times as many. In fact, there are over 150,000 hospitals *outside* the U.S. So, we may see some international firms "tunneling" into the U.S. markets via acquisitions, marketing-rights deals, etc., then pursuing the large, non-U.S. opportunity. Note, for example, the Thyssen-Bornemisza Group bought out Continental Healthcare Systems to add to its existing investments in the library automation markets (CLSI, BRS, BRS/Saunders - providing medical database services, among others).

Some Final Thoughts

The appearance of the same companies in segment after segment underscores the effects of maturity quite well. We expect to see continued consolidation of both segments and players within Health Care Computing, and more emphasis on *total solutions* from very few vendors. The battle lines will be drawn between the total-solution vendors (with a "we can satisfy all of your needs" marketing strategy, attempting to replace individual, entrenched systems at the customer site), and, the technologically-superior, individual-function, systems vendor, in league with established systems integrators, urging the customer to buy the very best product for each of its functions, then "connect" them (e.g., with LANs).

By focusing on historic performance and the size of U.S. spending on health care, without considering recent regulatory changes, traditional strategic analysis would have pictured health care computing as a wide-open, high-flying industry. This is clearly not the case.

The future appears to belong to the existing, well-established players who either offer a comprehensive solution or have the ability to integrate one. We would look among the best of the former for survivors, and among the latter for major new entrants.

7

Market Study #3:
Professional Services

7

Market Study #3: Professional Services

Introduction

The professional services sector of the computer services industry is in a period of rapid transition. As we see it, there are six major forces driving this change:

1. An easing of the long–standing shortage of basic programming talent;

2. The commoditization of the temporary personnel business, whose participants are pejoratively described as "body shops";

3. The consolidation of smaller players into powerful national and international firms, and the entry of new and very large players;

4. Historically low margins in traditional fee–for–service personnel services;

5. Increased customer demand for specialized talent, e.g., in DB2, in expert systems, in specific vertical market applications or software products;

6. The emergence of a newly perceived customer need for a single project manager and systems integrator, especially for those larger, more critical, front–office systems.

These are, of course, related forces. The gradual easing of the long–standing shortage of programming talent has reduced the demand for hourly COBOL programmers. This supply/demand equation has not yet reached equilibrium, and prices are, as a result, under considerable competitive pressure. For example, one major professional services firm's CEO reported that a farm equipment manufacturer was offering up to 1,000 "wholesale" COBOL programmers (his temporarily unneeded and costly staff) at $18 per hour, well below the actual cost of such personnel.

The Big–Eight accounting firms, lead by Arthur Andersen, have dramatically increased their penetration into the traditional markets for programming talent.   According to Input, the market research firm, accounting and management consulting firms now control 18% of the $8.7 billion market for commercial professional services.   This penetration by The Big Eight promises to continue thanks to their high–level contacts, nationally–known names, and early warning systems for prospect identification (their tax and audit relationships).

If this isn't enough, the big players have been getting bigger. Independents like Computer Task Group (CTG) and AGS Computers have grown rapidly through acquisitions, both in the U.S. and in international markets (CTG recently acquired Canadian and U.K. firms), while large, international firms have taken aim at the U.S. markets (e.g., CAP Gemini Sogeti's acquisition of DASD, Inc. and, more recently, CGA Computer Associates).

The result has been a competitive dogfight, especially in the major metropolitan markets.   Professional service "occupancy rates" (the percentage of employee time that is billed to customers) has been dropping, and traditionally low after–tax margins are getting even worse.   The hanging of "systems integrator" shingles by truly giant entities like GM/EDS, Martin Marietta, and IBM (see Chapter 8 for a thorough look at this market segment) has only exacerbated the competitive pressure and paranoia in the marketplace.

We believe the problem is not merely a temporary one related to the general slump in computer services and short–term market consolidations, and that the paranoia is justified. This conclusion clearly warrants an in–depth analysis of the market dislocations taking place and the strategic options open to the competitors.

Segmenting the Professional Services Marketplace

Figure 7.1 is our now–familiar three–dimensional market segmentation methodology, breaking down the professional services market sector into distinct market segments. While the full presentation of this tree diagram extends out many more levels, the major market segments can be clearly identified.

The "problem" market segment discussed above, providing programming talent at hourly and daily rates, involves the sale of software programming skills, primarily COBOL (the technology/ specialty), for systems development and implementation (the function), to information systems management and operations personnel (the customers).   To the extent that the traditional players have broadened their offerings, they have typically added technical skills in additional hardware environments or programming languages, while attempting to move "upscale" into systems analysis, design, and project management, with limited success.

We believe that this traditional market for COBOL (and other) programming skills, even with the addition of upscale analysis and project management talents, will continue to be under extreme competitive (and hence margin) pressure for the foreseeable future. Our belief is based on the expectation of a continued disequilibrium in the programmer supply/demand curve and, of equal importance, continued budgetary pressure on the traditional customer for such services, the information systems manager.

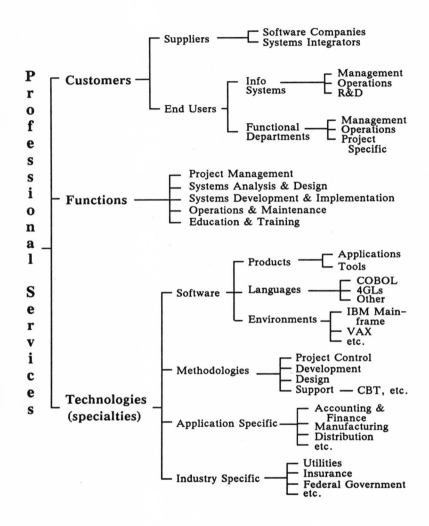

Figure 7.1

Our view of the other professional service market segments is far rosier, however. As there are over 1,000 market segments illustrated on Figure 7.1 alone, a very compressed view, let's restrict our examination of these other segments to major sectors (for newcomers, a "market sector" is composed of a market segment and relevant nearby segments). As we see it, the fifteen major sectors are:

	Customers	Specialties	Functions
1.	Info. systems	Software	Analysis & Design
2.	Info. systems	Software	Implementation
3.	Info. systems	Software	Project Management
4.	Info. systems	Environment	Operations
5.	Info. systems	Applications	Implementation
6.	Info. systems	Methodology	Implementation
7.	Suppliers	Software	Implementation
8.	Suppliers	Application	Implementation
9.	Suppliers	Industry Specific	Implementation
10.	End user dept's	Applications	Project Management
11.	End user dept's	Applications	Analysis & Design
12.	End user dept's	Applications	Implementation
13.	End user dept's	Methodology	Implementation
14.	End user dept's	Educate & Train	Project Management
15.	End user dept's	Industry Specific	Project Management

This overly simplistic view of myriad individual market segments allows us to examine strategic alternatives in an intelligible fashion. Before we do, however, some notes on our list of fifteen sectors are in order.

The first two sectors listed are the traditional "body shop" marketplace, the segments we find least attractive for the future. The third sector, complete software project management, has promise, but only in very large commercial accounts and government installations which are increasingly dominated by the large-scale system integrators (e.g., Computer Sciences, EDS, and now IBM). The fourth sector (traditionally called facilities management) has also gravitated toward the large operator (e.g., once again EDS). Sectors five and six, the provision of application or methodology skills to information services departments, hold great promise, possibly the only true growth markets remaining within the traditional professional services customer set.

Sectors six through nine represent the rapidly growing "OEM" market for professional services, i.e., the subcontracting of software, application, or industry–specific talents to a systems integrator or to a systems or software supplier. If you believe IBM, they would like to increasingly recruit outside providers of professional services as subcontractors.

The last six sectors are where the action is, in end–user departments. This is where we feel the greatest growth opportunities lie, but also where the skills of the traditional professional service firms have been weakest. Accordingly, let's examine the full range of strategic alternatives.

Strategic Alternatives

To summarize our thesis, the traditional professional services firm has watched as its primary product, skilled programmer talent, has become more and more of a commodity. Simultaneously, the budgets of their primary customers have been squeezed by increasing expenditures on software products, by major resource allocations to the maintenance of old systems, and by successful claims by user departments on discretionary system development funds. Last, but by no means least, new and more powerful competition has entered the market by "tunneling" from other positions of power, e.g., Big–Eight accounting firms whose traditional power base was in audit and tax, software products firms who have added after–sale support services, and now IBM.

We conclude that the traditional firms have four different paths open to them:

    1. Compete head–to–head for the commodity side of the business by becoming the low–cost provider;

2. Specialize in specific niches, e.g., applications, software product implementation and support, or methodologies;

3. Move laterally and sell to new customers, i.e., end users;

4. Join forces with, or supply services to, the growing set of powerhouse systems integration companies.

Many firms are trying to pursue the first option. Input interviewed information systems managers and discovered that vendor size and staff availability were the most significant reasons given in considering a professional services vendor; hence the rush by regional firms to grow into national and international firms, mainly through acquisitions.

Not everyone can be an internationally recognized low–cost provider of raw manpower, however. This first option is far-fetched for any not–already–large player; there is room for three or four very large firms that can use their very size to dominate the bidding for large–scale manning requirements, but not dozens. There is, however, plenty of room for smaller and higher-cost providers to exercise the second through fourth options listed above. Let's examine each.

The most suspect of the three remaining options is to become a subcontractor to a systems integrator or software products company. True, the temptation is great. Marketing costs are reduced or eliminated, and the environment is more stable than in multi–vendor, user–controlled projects. The risks are extreme, however, and only by spreading these risks over multiple partners would a supplier of subcontracted services avoid the possibility of potentially fatal dependencies on a current or future competitor.

Accordingly, we narrow in on two viable strategies for the small to medium–sized professional services firm: specialize and/or move into end–user services, preferably both. Specialization, referring back to Figure 7.1, could be in applications, industry knowledge, or methodologies. An example of the first, applications, would be in one or more software products, such as general accounting systems or database technologies, which require massive amounts of implementation assistance. An example of the second, industry specialization, would be an in-depth knowledge of the needs of wholesale grocers, and the development of a reputation as the premier provider of custom development services to that market. An example of the third, methodologies, would be the provision of computer-aided systems engineering (CASE), project management, or computer-based training methodologies as an extension of the general professional skills offered.

Regardless of specialty or market niche, taking on the full responsibility for all software, hardware, and communications, and for the integration of each, is an emerging opportunity for all professional services firms (and others) that we will examine in depth in Chapter 8.

It is useful to note that, increasingly, the money for large systems development projects rests with the end user. In large enterprises, the project managers for new user systems are more often than not selected from the user (versus data processing) community. In large projects, these tend to be up-and-coming managers, well respected (and well compensated) by their companies, who can compete as equals or superiors with the information systems manager for funds and management attention. What they want to buy, however, is decidedly *not* COBOL programming talent. They want to buy solutions, including project management, design services (not program design, but business system design), communication interfaces to the data processing community (i.e., translations of user needs into data processing terms and vice versa), training, and support services.

The addition of complete project skills to a vertical market or application specialty would allow a professional services firm to develop account control within its targeted set of customers. Accordingly, we believe that specialization in a specific vertical market segment or application, followed by an expansion of services to fully serve that unique market, is a winning strategy for the smaller firm.   As an example of this approach, a professional services firm might first examine its existing customer base to determine one or more potential areas of specialization, and do a competitive analysis to see what threats or opportunities exist if it were to specialize in those markets.   Secondly, the additional areas of competence required to properly service that market, from both an information systems and user standpoint, would be itemized to determine feasibility; can these skills be developed or acquired?   The last step would be to execute the strategy by acquiring the people and skills (or companies) that complete the market needs and then market aggressively to that segment, perhaps even becoming the full–scale systems integrator in a market niche.

Conclusions and Predictions

Simply put, there are only two approaches open to the plain–vanilla professional services marketers: become the lowest–cost provider or specialize. The first approach is only open to the top firms who can truly become all things to all people, and this will surely exclude some players who currently think they can. The biggest will continue to grow bigger (primarily via acquisitions) in order to drive marketing costs down and occupancy rates up by bidding on larger and larger contracts. In the second case, specialization, mergers and acquisitions will also play a very significant role. To specialize in cross–industry applications, professional service firms will acquire software companies or will themselves be acquired to service after–sale markets. To specialize in methodologies, professional services companies will acquire training and support companies, CASE software companies, and structured design firms. Specialization in vertical markets will see consulting firms, professional service companies, and software providers join forces, often via merger.

Typical professional services projects will grow from the low-end of six figures to the top-end of seven figures, but they will change in character. Formal proposal requests and fixed-fee contracts will be the norm in commercial projects, not the exception, as today. Complete project services, from detailed business requirement analyses through end-user support will be required from the prime contractor, either directly or through subcontracts.

Successful medium-sized firms will stake out territories in specialty markets (e.g., computer-aided design, human resource systems, bank back-office processing), and many will then be consolidated into the larger systems integrators. New firms, some not currently in computer services, will make major acquisitions to enter the systems integration/large project fray. Likely entrants are the Regional Bells, Ford and/or Chrysler, and the major aerospace firms.

Having established the basis, let's turn to the exciting market for systems integration.

8

Market Study #4:
Systems Integration

8

Market Study #4: Systems Integration

Introduction

"Systems integration" has surfaced as one of the emerging trends in the information industry. Market researchers, industry analysts, and hardware and software vendors are touting this concept as the opportunity of the 1990s, when corporate users will demand that their software, hardware, and communications needs be satisfied by a single vendor, at the lowest cost, with full compatibility and interoperability. In this perfectly integrated world, no longer will users have to string together lines of COBOL code and yards of coaxial cable in order to move a document or a spreadsheet from desktop to desktop. The systems integrator will deliver a fully functional, customized solution to cure the manager's daily headaches. Systems integration, the vendors seem to be saying, will replace piecemeal systems development with a total, end–to–end computing/communications infrastructure.

Definitions

How, then, should this new concept be classified? A number of industry observers have offered definitions, all of which are useful when taken in context. The August, 1986, issue of the *Tech Street Journal* pondered the issue of systems integration and offered the following:

> *"Something's afoot. If you think about it, there are dozens of reasons 'systems integration' ought to be on a vendor's list of major market opportunities, and only one*

reason why it shouldn't. The latter, of course, is that it can't be done.

"Well, it CAN be done, but at a stiff price. At a simple level, systems integration means cobbling software and hardware together so one vendor's LAN will interface to another's through a third vendor's PBX via data links running on another vendor's packet switching service tying together terminals running under somebody's SNA, and so on. That might even be possible. Electrons move so fast that communications overhead may even be tolerable in select instances."

Vendors performing systems integration know that the "stiff price" mentioned above cuts both ways. It refers not only to the size of the user's bill for a systems integration project, but to vendors who are losing money by running over budget and not succeeding in their efforts to deliver total systems.   "Cobbling hardware and software together" is a simple way to define the process of systems integration, but it says little about the business of being a systems integrator.

A more concise definition was offered by Hambrecht and Quist when describing the business of SHL Systemhouse, an Ottawa-based systems integrator servicing the Canadian and U.S. governments.   This public company is possibly the only pure play in systems integration.   According to Hambrecht and Quist:

"The systems integrator typically sells a solution (e.g., an office automation system), for a fixed-price contract, and assumes full responsibility for the delivery of the required function on time. The benefit to the user is a) vendor assumes the risk of cost over-runs; b) user (sic, we think they meant to say vendor, Ed.) provides necessary hardware and communications equipment and skills; c) vendor assumes the risk of systems failure."

The fixed-price aspect is an important factor that distinguishes systems integration from contract programming.   "Body shops" typically charge users on a per–diem, cost–plus basis, for the services of a programmer. A systems integrator, on the other hand, accepts total project responsibility for delivering a hardware, software, and/or communications solution for a predetermined price. (Strictly speaking, some systems integration projects in the federal government can include cost–plus elements. In the case of SHL, their reported revenues include hardware resale pass–throughs that may be part of their fixed–price projects.)

This notion of "total project responsibility" is another of the key distinguishing factors of systems integration relative to other computer–based professional services. Successful systems integrators must possess enough bidding and project management expertise to estimate projects accurately and deliver a price that contains a reasonable profit. Contractual risk in a systems integration project is high, and the skills required to develop a fixed–price bid are not trivial. Once this barrier to entry is overcome, however, systems integrators can realize at least twice the 5–10% pretax profits of pure "time and materials" consulting organizations.

To date, the largest customer of fixed–price systems integration has been the federal government, which (leaving certain well-publicized overruns aside) is one of the savviest purchasing organizations on the face of the earth. Several companies have mastered the art of marketing systems integration to the Feds, including the aerospace and defense contractors, professional services firms, and divisions of the major computer and communications firms. Martin Marietta, Computer Sciences Corporation, Boeing, EDS, Logicon, Planning Research Corporation, CACI, and SAIC are just some of these companies, all having major presence in the Washington, D.C. Beltway area.

Actually, the federal government is made up of several large "customers" (the individual agencies), each with its own computing and communications needs. One marketing strategy of federal systems integrators is to build a track record in a particular agency, and then use that acquired expertise to bid on projects in other selected agencies. Segmenting the federal government into its constituent agenices is a formidable task, and we will not perform that exercise here. In order to complete our definition of systems integration, however, we will look at the major market sectors and break them down using our three–dimensional methodology.

Segmenting the Market

Our definition of systems integration accepts the concepts of fixed–price bids and total project responsibility as a starting point. We would expand the definition to include all of the relevant functions (hardware procurement, software development, network construction), customers (government, commercial, and affiliated groups), and technologies (both DP and communications). Figure 8.1 shows our definition of systems integration by using our three–dimensional segmentation methodology.

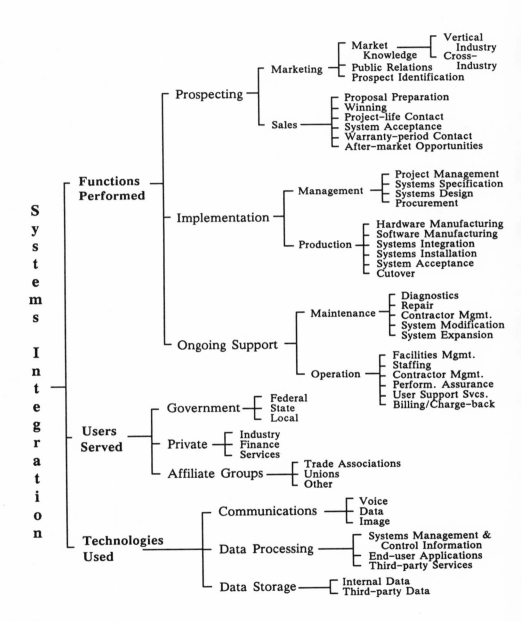

Figure 8.1

We have broken down the systems integration world into six major market sectors, each having many segments which would require detailed examination before competitive conclusions could be drawn; however, it is useful to discuss the general characteristics of each sector. Since systems integration can cover a multitude of projects of different sizes, some of the sectors are composed of projects which may also be components of larger systems. That is, private communications networks contain all the elements of both private voice networks and private data networks, data communications systems contain the elements of both private data networks and data processing systems, and fully integrated systems consist of all of the above. Figure 8.2 shows these relationships.

Market Sectors for Systems Integration

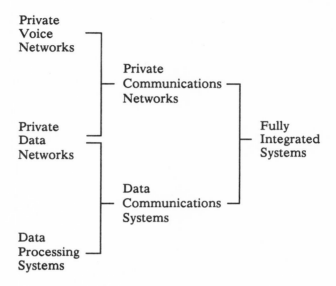

Figure 8.2

Private Voice Networks

The deregulation of the telephone company produced the opportunity to provide integrated systems of dedicated voice facilities for private (i.e., non common–carrier) use. Usually, the system integrator uses common carrier facilities for transmission, but bypass is also a possibility. This sector is heavily weighted toward the hardware, with PBXs, tandem switches, multiplexors, cables, handsets, etc., being the greatest part of the sales dollars. Network control and management is the major value–added component added by the system integrator. Figure 8.3 shows the characteristics of the major segments in this sector. This sector is still in its infancy, but the telephone companies, the PBX manufacturers, and some of the larger federal government systems integrators are all moving into the game, with joint ventures being frequently seen for the larger projects.

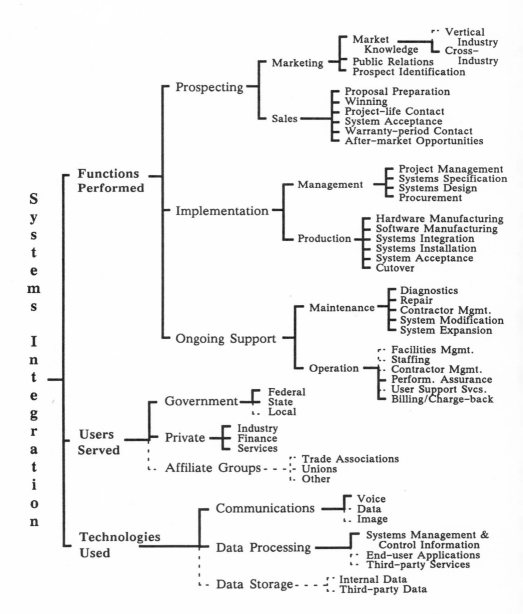

Figure 8.3   The Private Voice Networks Sector

Private Data Networks

This sector was broken into its three major segments and ana-
lyzed in Chapter 4, so we won't repeat that discussion. The usual
characteristics of the segments in this sector are highlighted in
Figure 8.4.

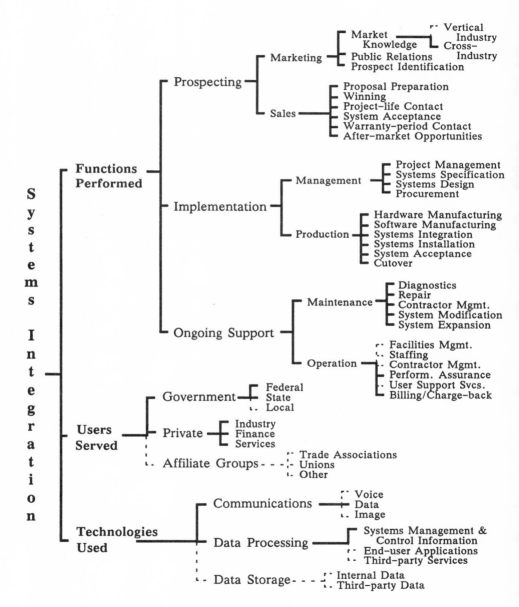

Figure 8.4   The Private Data Networks Sector

Data Processing Systems

The most mature systems integration sector is certainly this one. The major functions are software oriented, including the development of new applications, the customizing of existing applications, and, most importantly, the integrating and interfacing of different application modules. An example might be the development of a labor cost distribution system to be integrated with an off–the–shelf payroll module, interfaced to the benefits system, and integrated into the project management system through customization of the project control application. The players in this sector are generally the software professional services companies. Figure 8.5 shows the primary characteristics of the segments within this sector.

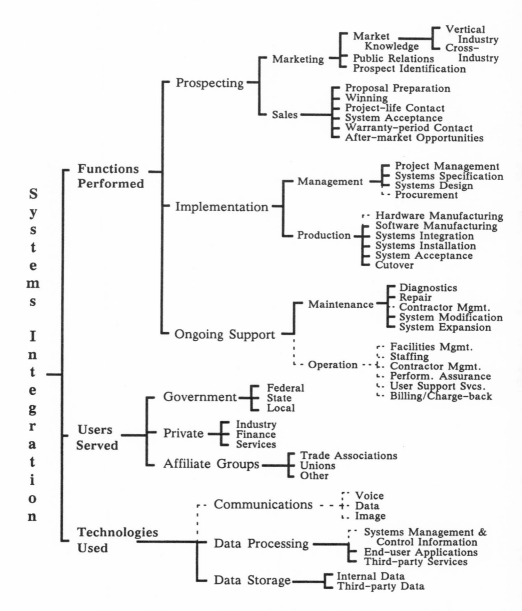

Figure 8.5   **The Data Processing Systems Sector**

Private Communications Networks

Combining voice, data, and other communications into a commonly–managed and controlled network requires the skills of
both the voice and data network sectors.   Since few (if any) suppliers are competent in both voice and data, the large federal
government systems integrators are the lead entrants in this
emerging market sector.   This will certainly change in favor of
the classical telephone equipment suppliers as the Integrated
Services Digital Network (ISDN) becomes widespread reality in
the next decade.   Figure 8.6 portrays the dominant segments of
this sector.

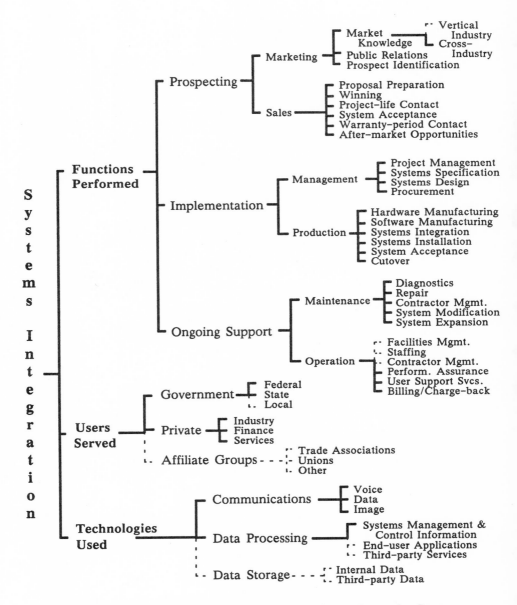

Figure 8.6   **The Private Communications Networks Sector**

Data Communications Systems

Combining the skills of data networking and data processing is necessary for the integration of communications–intensive information systems. For example, the furnishing of an integrated manufacturing control system, including material requirements planning, inventory control, order entry, etc., which ties together several plants and ordering points via a data network and interfaces with existing administrative and financial systems would be an application in this sector. This sector is the largest of the six and is growing rapidly. Most of the "mission critical" systems expected to be developed in the next several years will fall into this sector. Many segments are encompassed, both cross industry and vertical industry. The federal government is still the largest single customer (but this one customer is, in truth, many); however, private industry and state and local government represent the highest growth opportunities. Figure 8.7 shows the characteristics of the segments in this sector.

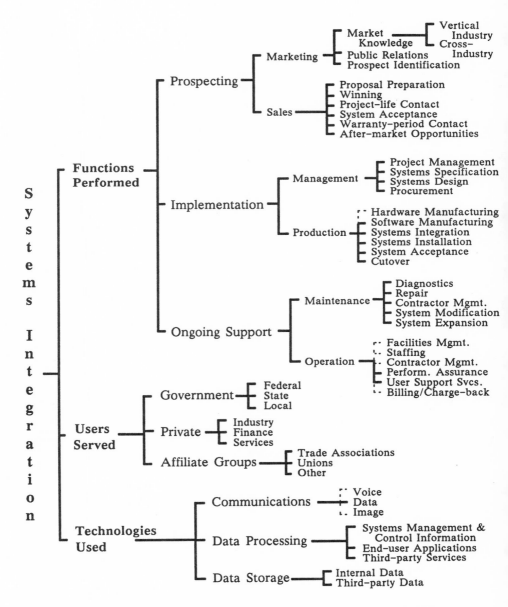

Figure 8.7   **The Data Communications Systems Sector**

Fully Integrated Systems

This sector exists mostly in the dreams of the large systems integrators; nevertheless, some of these dreams have come true.
EDS was purchased by GM so it could fully integrate the various
voice, data, and information systems of GM. Boeing has given
its Boeing Computer Services some similar projects. The federal
government is starting to include voice communications requirements in its systems requirements. It remains to be seen if this
market sector grows or if Private Communications Networks (in
the form of ISDN technology) and Data Communications Systems (using the ISDN networks) continue side–by–side. We
think it will be the latter for the foreseeable future. Since this
sector encompasses everything, Figure 8.1 is what it looks like.

Strategies for Success in Systems Integration: Introduction

The most attractive strategy for systems integration vendors will
be to pursue communications–intensive systems development for
commercial clients. We believe this opportunity will be enor-
mous. The dollar value of these projects will be in the $1 to $15
million range, with an occasional larger, multi–year contract being let to a consortium of vendors led by a large systems integrator. Many projects will involve primarily front office automation,
and many will be sponsored and paid for by line departments
other than DP/MIS.

In order to market commercial systems integration effectively,
vendors must possess expert knowledge of particular vertical industries. For example, expert knowledge in banking means more
than simply offering COBOL programming services to banks; it
may mean understanding the information needs of mid–sized
banks participating in the secondary mortgage market. Although
technical competence is certainly a critical element for an integrator, vertical industry knowledge will probably be the key pur-

chase decision variable, especially when the decision maker is a non–technical manager. Given this assumption, vertically oriented professional services firms are best positioned to grab the early lead in commercial systems integration.

In the near term, we believe federal systems integrators will have great difficulty trying to sell large–dollar–value (over $25 million) contracts to commercial customers. Their lack of vertical industry knowledge will be a major stumbling block, as will be proving the return–on–investment case for such major capital expenditures. In order to gain a toehold in Fortune 1000 accounts, these vendors must demonstrate industry knowledge and the ability to deliver systems that yield measurable cost reduction or revenue enhancement.

The detailed breakout for the data communications systems sector, which we feel contains the key opportunities for commercial systems integration, is presented in Figure 8.7.

Until recently, systems integration had been sold almost exclusively to federal government customers. Realizing that impending budgetary cutbacks may limit future contract awards, several federal systems integrators are now trying to attract commercial sector clients by emphasizing their experience in designing and managing large–scale data systems and communications networks. As an example, Martin Marietta has moved aggressively in this direction; it recently organized a project office dedicated to commercial systems integration.

This movement into commercial systems integration has not gone unnoticed. Veteran computer industry watcher Ulric Weil feels that if this shift continues vigorously, there may be major implications:

> *"System integration – customized design, implementation and ongoing support of complex computer–based applications – already so popular in the government, is gaining favor in the private sector. The increasing participa-*

*tion of the major aerospace vendors, many professional
services firms, and leading MIS users in industry and
business represent a potential threat to the traditional
computer companies."*

The long–term threat to computer companies is real. In a systems integration project, hardware is relegated to commodity
status, and the value added comes from the ingenuity of the system design, the ability to source computers, software components, communications hardware, and transport services at the
lowest cost, and the management systems that control this complicated process. In order to implement these functions, e.g.,
network topology analysis, workflow design, software engineering, and subcontractor management, the integrator must have access to a competent team of technical professionals.

The implication is that computer manufacturers who want to succeed as commercial systems integrators must build, buy, or align
themselves with a high–quality, vertically–oriented professional
services organization. This notion strikes fear into the hearts of
the CEOs of the capital–intensive computer manufacturers, who
have historically tried to avoid personnel–intensive activities such
as professional services. In the face of sagging worldwide demand for undifferentiated hardware and user demands for industry–specific integrated solutions, however, IBM and the other
computer vendors will have to consider seriously the risks of *not*
hanging the systems integration shingle as part of their product
portfolio.

IBM's Role

IBM is taking a two–pronged approach to commercial systems
integration. Its Complex Systems organization, under the Federal Systems Division, specializes in designing large transaction-
based systems for selected commercial customers, currently including United Airlines and Hospital Corp. of America. In addition, most of IBM's recently announced SolutionPacs have a sys-

tems integration approach, albeit on a limited scale, for the commercial customer. (Interestingly enough, both of these initiatives reside in what we refer to as the data communications systems sector: combining data processing and data networking skills.)

Although IBM can usually dominate any market it chooses, we suspect that it will tread lightly into commercial systems integration at this early stage of market development. IBM's sales organization is masterful at "moving iron" and protecting its installed hardware, but is relatively inexperienced at bidding fixed-price, industry-specific solutions. This deficiency will be rectified gradually at the low end: IBM's SolutionPacs announcement is a clear but cautious statement of intent to strengthen the company's capabilities in delivering integrated solutions. At the high end, IBM's Complex Systems organization will concentrate on cross-industry back-office automation for a few customers, without a vertical market emphasis.

We believe this cautious posture by IBM will leave numerous commercial needs unsatisfied over the next few years. Given this assumption, which vendors are best positioned to capitalize on these opportunities? In order to answer this question meaningfully, we developed a framework for evaluating the relative strengths of computer software and services vendors against the critical skills required for success in commercial systems integration.

In our analysis, we subjectively rated a sample 20 companies and separated them into four major groups. The objective of this exercise was to identify the operating fundamentals of a systems integrator and to locate companies that possess the optimal mix of these fundamentals and vertical market expertise. A brief summary of our major findings is found in the following paragraphs:

The Role of Federal Contractors

Because of their leading–edge technical skills and their project management experience, federal contractors are best qualified to undertake complex systems integration projects.   Their willingness to assume total project responsibility will be welcomed by commercial customers––if they can get their feet in the doors of corporate America.   As they are currently organized, federal contractors have neither the vertical industry knowledge nor the sales organizations to attract the interest of Fortune 1000 customers.   In order to improve their visibility in commercial accounts, the federal contractors must seek strategic alliances with commercial software and services vendors or outright acquisition of vertically–oriented professional services firms.

The Role of Software Vendors

In order to succeed at marketing off–the–shelf solutions, software vendors must have a great degree of specialized knowledge about the markets they serve.   Their operating leverage resides in strong distribution channels, tight quality control and product reliability, and packaging skills.   This emphasis on pre–packaged solutions, coupled with a lack of project management experience, probably relegates software vendors to the role of subcontractor in commercial systems integration projects.   For example, Arthur Andersen has agreements to use the products of independent software vendors Management Science America and McCormack & Dodge in its commercial systems integration efforts.   To break free from this second–class status, software vendors must partner with or build a professional services organization to add customized code around their products.   However, any such shift would be a major diversification for these firms.

The Role of Professional Services Companies

Within this group of companies, there is a wide spectrum of qualifications. Contract programming "body shops" lack both vertical knowledge and project management skills and would have a tough time restructuring their businesses to acquire these attributes. Custom software developers often possess vertical knowledge, but many are missing the total project responsibility ingredient. Only a handful of professional services firms possess all the relevant skills (vertical knowledge, project management, and experience with data communications), and we expect these firms will carve out a healthy slice of the early profits in commercial systems integration.

The Role of Consulting/Big Eight Firms

By virtue of their contacts with senior management, The Big-Eight accounting firms and the technology-oriented consultancies are well positioned to exercise considerable account control. Despite this marketing advantage, many of these firms lack some of the requisite basic skills. Often, they are reluctant to accept total project responsibility and fixed-price contracts, favoring a less risky time-and-materials billing schedule. Their expertise resides mainly in cross-industry problem-solving, such as financial control systems and order-entry functions, and not in industry-specific solutions. Of the companies in this group, Arthur Andersen has made some progress in assuming total project responsibility, and we expect others will follow their lead.

Other Players and Factors

This grouping of companies does not constitute an exhaustive list of potential competitors. As mentioned earlier, computer hardware manufacturers may take a stronger initiative in designing mission-critical systems for their preferred clients as a way to reinforce account control. For example, Unisys's SDC subsidiary is clearly positioned to tackle the commercial systems integration market. Aggressive MIS departments also offer a serious challenge, especially those accustomed to operating as profit centers. In fact, companies such as AMR (parent of American Airlines), Weyerhaeuser, and Boeing have spun out their MIS departments into separate businesses. These may be the real sleepers in commercial systems integration: MIS shops run by profit–driven managers who thirst for extra–company challenges.

In our ongoing discussions with end–users and MIS directors, we are convinced that demand for commercial systems integration is robust. Many vendors, on the other hand, have not clearly articulated their strategies for satisfying these very specific user needs. In the following paragraphs, we raise some of the major strategic issues that should be addressed by any potential commercial systems integration vendor.

Commercial systems integration is closely linked to the automation of strategically important business functions to sustain *unique* competitive advantage. This implies that a systems integrator may have to choose non–competitive clients when bidding for these key systems. For example, an integrator could design a state–of–the–art railcar maintenance and tracking system, complete with electronic data interchange capabilities transported via fiber links using track rights–of–way, for a company like Norfolk Southern. Should this same integrator then bid a similar system for that customer's archrival, CSX Corporation? Astute users may demand non–compete language from the integrator to insure against this situation.

As mentioned in Chapter 7, spending for large systems development projects is increasingly coming directly from end–user budgets, and not from the MIS department. In some companies, these projects are originated and paid for by line managers in non–MIS functions. This trend plays well to an aggressive vendor like EDS, who often bids around internal data center management when it proposes systems integration projects. In contrast, SHL Systemhouse often bids *with* data center management on smaller dollar–value projects. Should systems integrators sell directly to the CEO, to the top MIS person, or to line managers in functional positions? The successful vendors will be sensitive to these distinctions and will adjust their selling tactics appropriately, depending on the prospect's environment.

Despite our concerns, we see evidence that the move to mission–critical systems is genuine, and that "commercial systems integration" is fast becoming a well–worn phrase in the positioning statements of a host of information services vendors. The real winners in this arena, however, will be the vendors who can translate the buzzwords into action and deliver on the promises of this emerging market opportunity.

9

Market Study #5: Multiplexors

9

Market Study #5: Multiplexors

Introduction

This chapter will explore two related markets: the market for data/voice multiplexors and then the market for intelligent multiplexors.

Data/Voice Multiplexors

We think all the publicity about Integrated Services Digital Networks (ISDN) is producing something after all: user demand for some of the promised services, now. On the opposing side is the natural reluctance of the Regional Bell Operating Companies (RBOCs) to use their local loops more efficiently when doing so yields lower short–term revenues. A difficult conundrum for the RBOCs. Here's another example of technology bumping heads with invested interests. This article will explore the two sides of the issue of how long the invested interests will hold off the inexorable march of technology. We'll then look at some of the players and their prospects for success.

The Case for the User

The breakup of the Bell System reduced the cross–subsidization of local telephone service by long–distance service, and the dust is still settling. For the user, this has resulted in a rapid drop in long–distance costs and an even faster rise in local service costs. The business customer with both voice and data requirements has seen the economic emphasis shift rapidly from reducing transport costs, via someone like MCI, to controlling the spiraling distribution costs, via technology such as PBXs, T–1 concentrators, microwave bypass, etc.   Enter ISDN.

ISDN offers the user the opportunity to use a single circuit for several simultaneous functions as opposed to today's one–function–per–circuit environment. In a cost–based economic system, this would result in saving lots of money through increased utilization of circuit facilities. Sounds simple, doesn't it? Back to the real world.

The Case for the Telephone Companies

The Bell System breakup is often erroneously referred to as the *deregulation* of the telephone industry.   Don't believe it. AT&T's long distance revenues, which represent over 80% of the market, are still tariffed and regulated. Even worse, the regulation of local service has passed into the hands of local officials who tend to be rather sympathetic to the pleas of their local telephone company. How else can one explain the increasing net income of the RBOCs in spite of the poor results many have shown in non–regulated activities like customer premises equipment sales, computer stores, etc.? (Not that we don't believe that non–regulated activities will contribute significantly someday; they will, but for the time being the RBOCs are protected from the full financial consequences of the learning process by the high profits of local telephone service.)

The point is that, in the current environment, the operating telcos have little motivation to help users reduce local circuits through more efficient use. Their approach has been purely reactive to market pressures. For example, the new shared–circuit Centrex offerings would never receive the push they're getting if the private branch exchange (PBX) manufacturers hadn't started replacing the central office exchange local loop use (Centrex) with on–premises technology offering efficiencies and new functions.

Most domestic geographical areas have abundant local loop capacity; hence, combining functions to reduce demand is not a priority for a telco whose revenues and guaranteed profits are based on tariffed circuits in use. Combine this with the general unfamiliarity, and consequent discomfort, with anything having to do with data communications rather than voice, and you get reluctance in capital letters.

The Impact of Technology

Most of us have seen comparisons between the cost of computing and the cost of telecommunications over the last twenty years. The punch line is that if technology had impacted both equally, the phone call would be almost free today. We all know that isn't the case; however, divestiture and competition are beginning to have an impact. The introduction of devices that provide for multiple simultaneous use of a circuit is an example. ISDN will be the final form, but what can we do during the decade or more it's going to take for the entire telephone plant to be upgraded to ISDN? Try some data/voice multiplexing (DVM), also known as data over voice (DOV).

The technology to share the same old two–wire twisted pair of telephone wire for the simultaneous transmission of voice and data exists and is well–proven in service environments. In fact, for many applications the data transmission features and costs of DVMs are clearly superior to those of stand–alone modems, and the circuit is free! In practice, there is a small box at the user's end of the line which terminates the connectors, one for the data device (e.g., a PC), one for the telephone, and one for the line. At the other end could be another box, or, as in the case of a telephone company central office (CO), a rack-mounted card. One of the differences between ISDN and DVM is that, in ISDN, a common digital switch would handle both voice and data at the CO, but in a DVM environment the voice and data cables would terminate in different switches. Let's look at a couple of real examples for a better understanding.

One of the ways that the new Centrex is being implemented is by installing a data switch in the telco's CO next to the voice switch. AT&T calls their version of this a Datakit. In this configuration, the CO voice switch may be an older analog device, while the data switch is digital. The user has a DVM which multiplexes his data and voice to the CO on the same local loop.

At the CO, the telco splits the traffic to the two separate switches. These then route the signals either back to other on-premises users (the intercomm function if voice, LAN if data), or to external voice and data networks. Pretty slick, and selling well.

Another, less successful, example is telephone-company-provided packet switching. Most of the telcos installed packet switches in some COs during the last couple of years with the intention of picking off part of the increasing data traffic. By using DVMs, it is possible to use the local loops to collect the data without additional lines. Once at the CO, the data lines are split onto the packet switch which routes the packets to their destination, either local or through value-added network (VAN) carriers. Unfortunately for the packet switch and DVM manufacturers, the initial tariffing of this service has usually been designed to recover the revenue from displaced leased data lines rather than to promote the service. In other words, it was priced so high it didn't sell. Until an external threat such as bypass forces change, we expect this situation to continue. Talk about encouraging bypass!

The last, and most exciting, example doesn't exist yet due to continuing regulation. When the telcos are allowed to offer information services over their networks (and we believe this isn't far away), they will really get serious about using the local loop plant efficiently, because it will be their main competitive advantage. Imagine conversing with someone on the phone at the same time as you let your data terminal's fingers walk through the yellow pages to find a restaurant to meet at, or a phone number, or search any database. When the phone companies can generate revenues from such services (currently forbidden), the economics of local loop sharing will be less of a constraint. This will be a main feature of ISDN, but if the regulations are lifted before ISDN is commonplace (many years before, in our opinion), DVM is a good alternative. In fact, the widespread introduction of ISDN services may provide the largest market opportunity for DVM. By interfacing DVM–based ISDN–like services with installed ISDN equipment (a manageable software problem), the telcos will be able to provide end–to–end ISDN services over the entire network while phasing in the actual installation of ISDN switching equipment. This is a nice way to meet user demand without losing control of the capital replacement process.

The Players and Their Plays

We identified eleven players, which is a lot considering the total 1987 market looks to be around $30 million. In fact, there are two distinct market segments, the telco market and everyone else, and we estimate them to be about equal in size. Since the market is still very small and formative, we will treat it all as one segment for this analysis.

Applying our methodology for competitive analysis yields the relative competitive position (RCP) calculations shown as Table 9.1 and the matrix shown as Figure 9.1. We expect the market to grow around 25% in 1987 as compared with 1986, so the companies above the horizontal mid–line are projected to grow faster than 25% in this segment, thereby gaining market share. The ones doing the most business have high relative market share and are on the left side of the matrix. The larger companies are capable of bringing more muscle to the market (what we call

"applicable strength"), and this is represented by the size of the
bubbles.

DATA VOICE MULTIPLEXORS (DVM)

Company Name	Segment Revenue*	Segment Market Share	Relative Market Share
Coherent Communications	7.00	21.37%	1.00
Teltone	7.00	21.37%	1.00
Micom	6.50	19.85%	0.93
Applied Spectrum Tech.	5.00	15.27%	0.71
General Data Comm.	3.00	9.16%	0.43
Gandalf	1.00	3.05%	0.14
Infotron	1.00	3.05%	0.14
Racal–Milgo	1.00	3.05%	0.14
Seiscor	1.00	3.05%	0.14
Integrated Network Corp.	0.25	0.76%	0.04
Lear Siegler	0.25	0.76%	0.04
TOTALS	**32.75**	**100.00%**	

Company Name	Applicable Strength*	Circle Radius	Growth Rate*	Relative Growth Rate
Coherent Communications	10.00	1.8	40.0%	14.0%
Teltone	15.00	2.2	16.7%	-9.3%
Micom	25.00	2.8	30.0%	4.0%
Applied Spectrum Tech.	5.00	1.3	42.9%	16.9%
General Data Comm.	25.00	2.8	50.0%	24.0%
Gandalf	25.00	2.8	25.0%	-1.0%
Infotron	25.00	2.8	17.6%	-8.3%
Racal–Milgo	25.00	2.8	17.6%	-8.3%
Seiscor	25.00	2.8	-33.3%	-59.3%
Integrated Network Corp.	0.50	0.4	66.7%	40.7%
Lear Siegler	25.00	2.8	66.7%	40.7%

Segment Growth Rate   26.0%*
Start Year   1986
End Year   1987

*Vanguard estimates based on publicly available information, private
interviews, and market research. No representation is made as to
accuracy.

Table 9.1

RELATIVE COMPETITIVE POSITION
Data/Voice Multiplexors

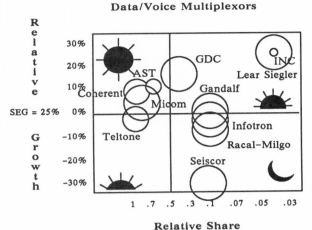

Figure 9.1

Applied Spectrum Technologies, Inc.--AST is a small company dedicated to the DVM market, so far. They have one of the newest technologies and are seeing success in both telco and non-telco applications. Siemens (formerly Databit) sells AST DVMs with their packet switches. AST is trying to penetrate the telco Centrex market as an alternative DVM supplier for systems such as Datakit.

Coherent Communications Systems Corp.--Coherent is another small company, with 25% of their business coming from DVMs. Due to an exclusive OEM deal with AT&T who brands Coherent DVMs as AT&T Datakit components, they are the market leader in the telco Centrex application. If AT&T Datakit sales take off, Coherent could be a big winner in the short term; however, they have all of their eggs in the AT&T basket for now. Coherent needs to succeed in its recently established efforts at direct sales in order to be a viable long–term survivor. OEMing is a great short–term tactic, but not a long–term strategy.

Gandalf Data, Inc., Infotron Systems Corporation, Micom Systems, Inc., and Racal–Milgo––All four of these major data communications equipment manufacturers do most of their DVM business as part of larger product lines, and are not major suppliers to the telco market. Micom is the most successful in the commercial market segment because of the "drag" sales from its matrix switching products.

General DataComm, Inc.––An early player in the DVM market, GDC has a nice position with Bell Atlantic's Centrex offering. If they manage to hold on to this customer (who is expected to be the fastest growing Centrex telco in 1987), GDC could have nice growth in DVM. They also compete in the non–telco segment within their product lines.

Integrated Network Corp. (INC), and Lear Siegler, Inc.––Both of these companies are new entrants trying to offer better technology. It's too early to call these, but they represent the typical newcomer trying to gain share through innovation.

Seiscor Technologies, Inc.––A Raytheon subsidiary, Seiscor was an early leader in selling DVMs to NYNEX companies for packet switching. They are succeeding about as well as NYNEX is in this effort, i.e., not much. It could always turn around pretty fast, but we don't expect it unless the regulations change.

Teltone Corp.––Probably the first company to offer DVM, Teltone has had some limited telco sales, but relies mostly on the normal network of telecommunications equipment distributors for their revenues.

Summary and Conclusions

DVM is a market bottled up by the regulated telcos, but the technology should prevail in time. We don't think 1987 is the year of explosive growth, but 1988 could be.

When the telcos are freed from restraints on information processing services, they will probably push DVM and ISDN in parallel as competitive edges to sell processing services. Centrex is the only hope for rapid DVM growth this year, and Coherent, AST, and GDC would all profit from this.

The non–telco market will continue to develop as users attack the cost of local service in every way possible. In particular, systems integrators of telecommunications networks, such as Computer Sciences Corporation, Bolt Beranek & Newman, Electronic Data Systems, Boeing Computer Services, AT&T, etc., will surely discover DVM as an effective weapon against local loop costs. Data communications equipment manufacturers will continue to integrate DVM components into their system product offerings to capitalize on existing wiring wherever possible.

We're bullish on the mid–range opportunities for those companies who can wait out the next year or so. The market should be viable for another five to ten years before ISDN slams the window completely shut. That's plenty of time in a high–tech business. Lastly, a normal shakeout can be expected during the next couple of years. Some of the players will exit, some will be gobbled up by bigger players, and maybe one or two will make a splash in the initial public offering market. The RCP matrix for this market will certainly be more sparse and less confusing in several years.

Intelligent Multiplexors: Is There Life After (Near) Death?

Anyone reading the 1986 annual reports from the major multiplexor manufacturers has seen all of the euphemisms possible for describing awful. Such phrases as "disappointing sales," "not up to plan," and, above all, "the computer slump" seem to be the order of the day. And all of this is despite the statements that the data communications market is supposed to be the fastest–growing user of communications around. Just maybe there's something strategically interesting going on. It's worth a look, anyway.

We previously explored the data communications market (Chapter 5) including the three–dimensional segmentation reproduced here as Figure 9.2. The appropriate parts of the segmentation have been highlighted to home in on the multiplexor markets.

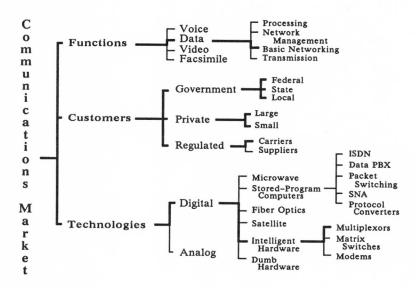

Figure 9.2

Based on our compilation of data from annual reports, market research publications, and our own database, we believe that the relative competitive position matrix for the multiplexor sector looks like Figure 9.3. As expected, the big boys, like Codex, Racal–Milgo, and Micom, are barely holding their own in a no-growth sector. Paradyne is straddling the border between setting sun and waning moon, but their difficulties, which have been well–publicized, extend beyond purely market considerations, and may now be over. But what's with Timeplex, General DataComm, and Infotron? The first two appear to still be on a rapid growth path, and Infotron is weathering the storm quite well for a competitor of small applicable strength. It makes one suspect that something is going on within the sector that isn't obvious at this point.

MULTIPLEXOR SECTOR

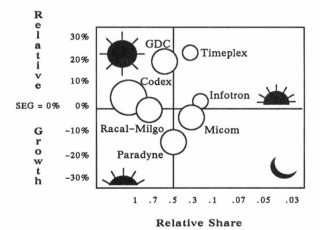

Figure 9.3

Using market growth projections from International Data Corporation's special report, *Communications Industry Review and Forecast*, July, 1986, we produced the sector growth–size matrix for 1986, shown as Figure 9.4.

MARKET SEGMENT COMPARISON
For Multiplexor Sector

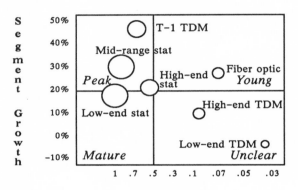

Figure 9.4

The actual results for 1986 were less rosy than IDC predicted (blame that old computer slump), but we think the relative growth positions of the different segments are valid. If anything, we suspect that the gap between the T–1 (1.54 megabits/sec.) Time Division Multiplexors (TDM) segment and the others is even wider than expected due to both faster growth in the T–1 segment and slower growth in the others. Anyway, let's see what it tells us.

In terms of maturity of market segments within the multiplexor sector, it appears that the young segments are fiber optics and T–1, with the latter being near a growth peak. The three statistical multiplexor segments are in transition between the peak and mature quadrants, and the older time division segments are well on their way out. Since this market sector is based mainly on the different technologies used to provide similar functions to a broad class of users, it is not surprising that the maturity of the segments is roughly equivalent to the age of the associated technology.

Now it's time to take a close look at the segment of most interest, the T–1 market segment. Our best information says that a relative competitive position matrix of this segment is as in Figure 9.5.

T–1 MULTIPLEXOR SEGMENT

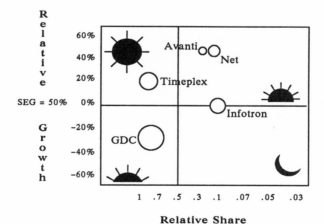

Figure 9.5

Timeplex is the noonday sun of this segment, GDC has fallen into the setting sun quadrant, Infotron appears to be moving into a rising sun position, and two newcomers to this particular sector—Avanti and Net—are healthy rising suns. (There are other players, but we have only shown the most important ones as of this date.)   Notice that the segment growth for this segment is 50%, so even GDC's −25% relative growth translates into real sales growth of 25%. GDC may be yielding market share in this segment, but it still represents growth for the company.

Assuming our data is reasonably solid, we should be able to return to the original relative competitive matrix for the multiplexor market sector and explain the apparent success of Timeplex, GDC, and, to a lesser extent, Infotron. It seems that the impact of being a major player in the high growth T−1 segment has been to allow these three companies to ride out the market slump with minimal reduction in growth of sales revenue. (What happened on the profits side is a different story, and depended more on cost containment in the down segments than revenue gains in the T−1 segment.)

Some of the strategic conclusions from this study are self-evident, and others aren't. It's pretty clear that being a rising, or, better yet, noonday sun in a young market segment is an attractive position. More subtle is the observation that being a setting sun in a young market even has advantages, at least for a while. Not being a player at all is clearly dangerous, especially for the big noonday suns of the mature segments.

Regarding the multiplexor market sector in particular, it didn't require our analysis for the major multiplexor manufacturers to figure this all out and start reacting. All now have T−1 products, and that segment is hotly competitive. We expect that a relative competitive position matrix for the T−1 segment produced next year would be quite different considering the large applicable strength of the newer players.

The T-1 market segment is one of the precursors of the Integrated Services Digital Network (ISDN). Fiber optic transmission and data-voice multiplexing are two others. What this all means for the communications market is that we are one step closer to what we call "the day of the almost-infinitely-available, almost-free digital bandwidth" (21st Century stuff, that). Someday the BOCs will drive those old copper local loops at T-1 rates on a routine basis, and everyone will have megabits available at their desks.

In the long run, we believe that those communications manufacturers who strategically position themselves to provide very high-speed facilities for integrated services will be the winners, and the T-1 segment is just the first example. It sounds simple, but it won't be because the major suppliers are in one of two camps. They either understand voice, or they understand data, but seldom do they do both well. IBM buys Rolm, AT&T tries the computer business, Wang buys Intecom, and the list goes on. These early positioning moves haven't had time to prove themselves, but they do represent the beginning of what we expect to be a long and difficult restructuring of the computer and communications businesses.

Bibliography

Abell, Derek F., *Defining the Business, The Starting Point of Strategic Planning*, Englewood Cliffs, New Jersey, Prentice–Hall, Inc., 1980.

Boston Consulting Group Staff, *Perspectives on Experience*, Boston, The Boston Consulting Group, 1968.

Davidow, William H., *Marketing High Technology*, New York, The Free Press, 1986.

Hax, Arnoldo C. (Editor), *Readings on Strategic Management*, Cambridge, Mass., Ballinger Publishing Company, 1984.

Henderson, Bruce D., *Henderson on Strategy*, Boston, The Boston Consulting Group, 1979.

Hofer, C. W. and Schendel, D., *Strategy Formulation: Analytical Concepts*, St. Paul, Minn., West Publishing Co., 1978.

Index

A

B

C

D